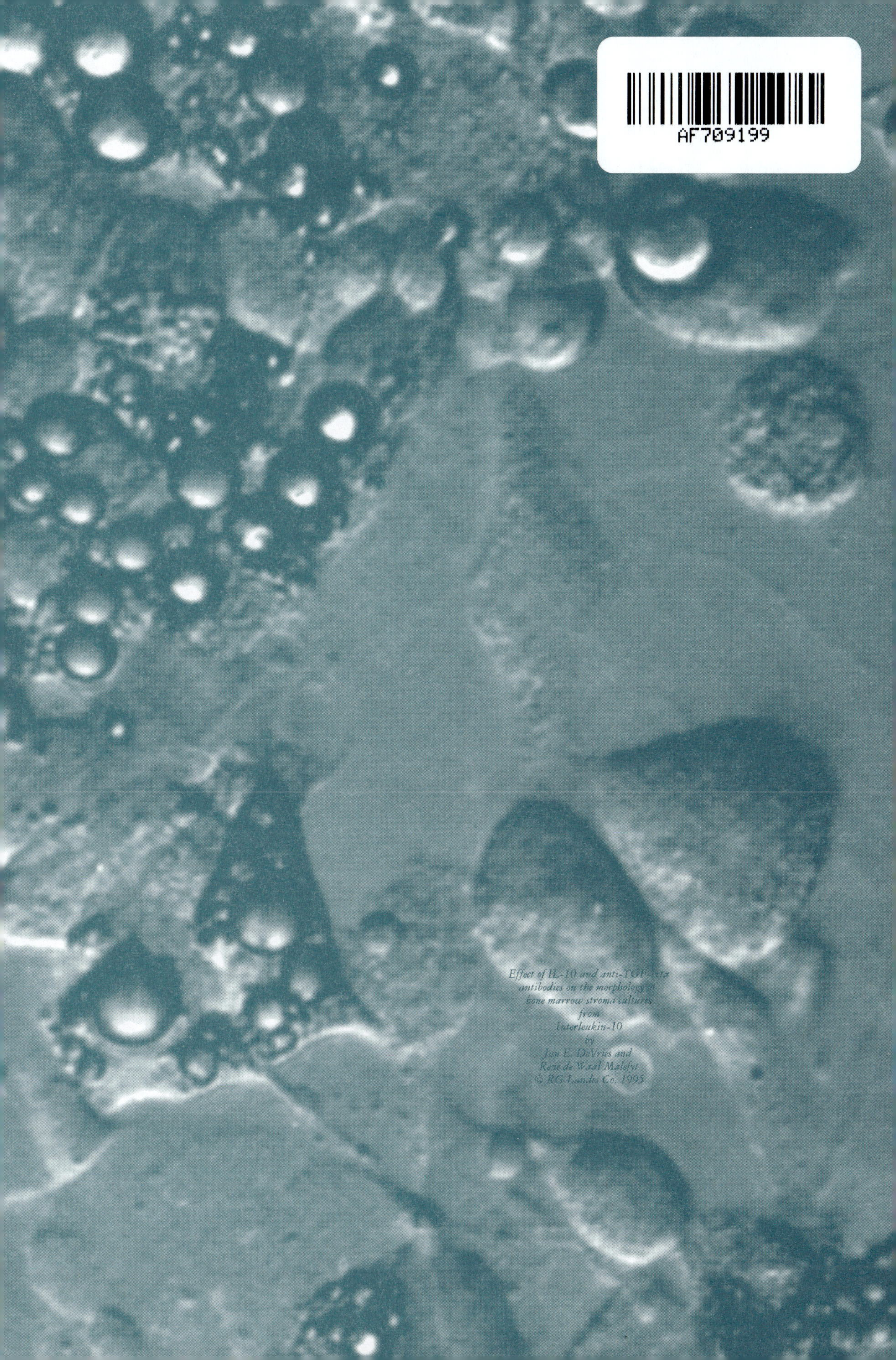

Effect of IL-10 and anti-TGF-beta antibodies on the morphology of bone marrow stroma cultures
from
Interleukin-10
by
Jan E. DeVries and René de Waal Malefyt
© RG Landes Co. 1995

Molecular
Biology
Intelligence
Unit

The Mammalian Molecular Clock

Simon Easteal
Australian National University
Canberra, Australia

Chris Collet
Australian National University
and CSIRO Division of Wildlife and Ecology
Canberra, Australia

David Betty
Australia National University
Canberra, Australia

R.G. Landes Company
Austin

MOLECULAR BIOLOGY INTELLIGENCE UNIT

THE MAMMALIAN MOLECULAR CLOCK

R.G. LANDES COMPANY
Austin, Texas, U.S.A.

Submitted: December 1994
Published: April 1995

Please address all inquiries to the Publisher:
R.G. Landes Company, 909 Pine Street, Georgetown, Texas, U.S.A. 78626
or
P.O. Box 4858, Austin, Texas, U.S.A. 78765
Phone: 512/ 863 7762; FAX: 512/ 863 0081

U.S. and Canada ISBN 1-57059-177-6

International ISBN 3-540-58902-3

Library of Congress Cataloging-in-Publication Data

Easteal, Simon, 1952-
The mammalian molecular clock / Simon Easteal, Christopher C. Collet, David Betty.
p. cm. — (Molecular biology intelligence unit)
Includes bibliographical references and index.
ISBN 1-57059-177-6 (hardcover : alk. paper)
1. Mutation (Biology) 2. Molecular evolution. 3. DNA—evolution.
4. Proteins—Evolution I. Collet, Christopher C., 1955- . II. Betty, David, 1953-.
III. Title. IV. Series.
[DNLM: 1. DNA—physiology. 2. Amino Acid Sequence—physiology.
3. Periodicity. 4. Evolution. 5. Mammals—physiology.
6. Genetics, Biochemical. QU 58.5 E13m 1995]
QH460.E27 1995
DNLM/DLC 95-6272
for Library of Congress CIP

Publisher's Note

R.G. Landes Company publishes five book series: *Medical Intelligence Unit, Molecular Biology Intelligence Unit, Neuroscience Intelligence Unit, Tissue Engineering Intelligence Unit* and *Biotechnology Intelligence Unit.* The authors of our books are acknowledged leaders in their fields and the topics are unique. Almost without exception, no other similar books exist on these topics.

Our goal is to publish books in important and rapidly changing areas of medicine for sophisticated researchers and clinicians. To achieve this goal, we have accelerated our publishing program to conform to the fast pace in which information grows in biomedical science. Most of our books are published within 90 to 120 days of receipt of the manuscript. We would like to thank our readers for their continuing interest and welcome any comments or suggestions they may have for future books.

Deborah Muir Molsberry
Publications Director
R.G. Landes Company

CONTENTS

PREFACE

While we were preparing this manuscript, three of the most important contributors to the ideas we set out to discuss, Motoo Kimura, Ernst Mayr and Linus Pauling, passed away. Linus Pauling, with Emile Zuckerkandl, coined the phrase "molecular clock". Together they were among the first to investigate rates of molecular evolution and developed the beginnings of a theoretical framework to explain evolutionary rate constancy. Ernst Mayr was a fierce critic of the molecular clock hypothesis. With his deep understanding of the complexity of adaptive evolution he expected that the evolution of DNA and proteins would reflect that complexity. By focusing attention on the discrepancy between the evolution of molecules and of phenotypes he helped stimulate an explanation for the discrepancy. The explanation was provided by Motoo Kimura in the form of the neutral theory of molecular evolution, which remains as one of the important ideas in the history of biology.

The legacy of these great scientists is the rich and active field of molecular evolution and a way of thinking that is increasingly permeating other areas of research. This book is a tribute to their contributions. We provide an historical context to current issues and a synthetic interpretation of data. Our conclusion is that the rate of DNA evolution does not vary among mammalian lineages. This conclusion, which in itself is controversial, has a number of even more controversial implications. By drawing attention to these we hope to stimulate further discussion and inquiry, and in so doing to honor the legacy of the pioneers of this field.

Molecular evolutionary studies are continually increasing in extent and complexity, largely as a result of technological advances. This trend will continue with the "big science" approach to genomic research. Increasing numbers and kinds of investigations are becoming possible as technological limitations are removed. In all of this it is important that data be collected in a way that addresses important issues. Our hope is that issues raised in this book will provide some focus for the future acquisition of comparative sequence data.

Part of this work began life as a component of D.B.'s Master of Arts thesis in the Department of Archaeology and Anthropology, Australian National University. The preparation of that thesis was greatly helped by guidance and advice from C. Groves. We are grateful to I. Jakobsen (ANU) for carefully reading, correcting and commenting on parts of the manuscript; to S. Smith (CSIRO) for assistance with the graphics; to I. Newman and B. Staples for their energetic assistance in the CSIRO library; and to S. Collet and P. Easteal for patience and understanding.

CHAPTER 1

INTRODUCTION

The molecular clock hypothesis is that DNA and proteins evolve at an approximately uniform rate. This is one of the most elegantly simple concepts in biology, but it is also one of the most contentious. Put forward nearly thirty years ago, it is disputed as much now as it was then. A molecular clock has enormous potential value in helping to understand both patterns and processes of evolutionary change and this gives it great appeal. Controversy has arisen, however, because acceptance of the molecular clock has presented challenges to a number of conventional ideas.

In this book we discuss the origin of the molecular clock hypothesis and the subsequent work aimed at testing it. We show how this work has been undertaken in the context of developing molecular biological techniques and methods of data analysis. We discuss how the molecular clock relates to theories of molecular evolution, and the implications of a molecular clock with respect to mechanisms of mutation and to patterns and processes of organismic evolution.

Our focus is on mammals. This reflects our own expertise, but it also reflects the fact that much of the analysis of molecular evolutionary rates has taken place in mammalian species. Because of this many of the most interesting findings resulting from molecular evolutionary rate studies have been in mammals, including humans.

The molecular clock hypothesis arose from the observation that some proteins appeared to accumulate amino acid changes at equivalent rates in different lineages. The sequence comparisons on which the proposed hypothesis was based were very limited by today's standards (discussed in chapter 2). The acquisition of amino acid sequence data for homologous proteins in many different species was a substantial feat but was not enough to do more than suggest a hypothesis. Testing that hypothesis has been possible with the continued accumulation of amino acid and, more recently, nucleotide sequence data, as well as direct genomic comparisons based on DNA reassociation kinetics (chapter 4).

There have also been developments in methods of comparative sequence analysis. Much of the dispute over the molecular clock has arisen from different conclusions being drawn from the application of different methods of analysis. These issues will be discussed in chapters 5 and 6.

The conceptual importance of the molecular clock hypothesis was more fully realized with the emergence of "the neutral theory of molecular evolution". The neutral theory is as elegantly simple as the molecular clock and it has been just as controversial. Neutral theory relates the process of mutation in the DNA of individuals to the establishment of mutations in species over evolutionary time. In doing so it brings together the fields of population genetics and molecular evolution.

The formulation of neutral theory represented a break from the consensus view achieved through the synthetic theory of evolution, which emphasized the importance of natural selection. Neutral theory was, thus, widely perceived as an important challenge to the synthetic theory. Its development has been inextricably linked with the molecular clock hypothesis and this association is discussed in chapter 3.

In its simplest formulation neutral theory predicts a molecular clock. However, when results have been interpreted as rejecting a molecular clock they have often led to a modification rather than a rejection of neutral theory. An early example of this involved the finding of disparate results between analysis of protein and DNA data which led to the development of what was referred to as the "nearly neutral theory of molecular evolution". This is discussed in chapter 4.

The molecular clock has important implications for understanding the pattern of evolution, and molecular data now play a major role in the estimation of evolutionary trees (phylogenetics) and on the systems of classification based on these (systematics). This is because molecular data provide quantitative estimates of genetic differentiation; this is not true of the plastic phenotypic, usually morphological, traits on which much of traditional systematics is based. We do not discuss molecular systematics in any detail here, but the issue of a molecular clock is relevant in two respects. First, the use of some methods of phylogenetic analysis depends on evolutionary rate constancy. Second, if molecular evolutionary rates are constant then divergence times of species can be derived from molecular data, even when fossil evidence is not available.

The potential of the molecular clock in this respect is reflected in its enthusiastic acceptance by some (but by no means all) systematists and paleontologists. For example Gould[1] considered that a general solution to problems of phylogeny had been found. If the clock could be set by dating a few evolutionary branch points with independently derived evidence from fossils or geography, then clear patterns of relationships between organisms would emerge with the continued accumulation of molecular data. The molecular clock provides a basis for understanding the interrelationships of the biotic diversity on this planet and of the evolutionary mechanism by which it arose. Here we discuss the impact of molecular data on our understanding of the pattern of mammalian and particularly primate evolution (chapter 10). In doing this we draw on information from paleontology and biogeography (chapters 7 to 9).

It is important to understand from the outset that a molecular clock is not analogous to a metronomic clock. The rate of molecular evolution in the simplest (neutral) model reflects the rate of mutation. Mutation is assumed to be a stochastic process and molecular evolutionary rate is thus expected to have an element of stochasticity, somewhat analogous to the process of radioactive decay. Thus over long periods of geological time rates appear approximately uniform, but over short periods or when short sequences are compared variation may be evident.

It is also important to understand that this book relates primarily to simple nonrepetitive DNA, including protein coding regions. The principal type of sequence change being considered is base substitution. There are of course many other ways in which changes can occur within genomes; these are undoubtedly important in evolution, but they are, for the most part, not relevant to the present discussion.

The dependence of DNA and protein evolution on the nature of mutational mechanisms means that evolutionary rate will be uniform only if mutation rate is uniform. We discuss the evidence that mutation rates are affected by differences in rates of germline cell division among species (chapters 4, 5 and 6) and between males and females (chapter 11) and by differences in metabolic rate (chapter 12).

The discussion of molecular evolutionary rates presented here overlaps some very diverse disciplines, including anthropology, human genetics, mammalogy, molecular biology, molecular evolutionary genetics, mutagenesis, paleontology, population genetics and systematics. It also involves a number of complex theoretical and technical issues. These include the mathematics that comprise the neutral theory and related areas of population genetics and molecular evolution, the technical procedures used to obtain the various kinds of molecular data, the models used in the estimation of the number of substitutions occurring between compared sequences, and various methods used in estimating phylogenetic relationships. It would be impossible to provide a comprehensive background in all of these. We have endeavored to present as simplified an account of these as is needed to obtain a clear understanding of principles. A more detailed coverage of many of these issues can be found elsewhere.[2-5] Similarly, our purpose is not to present a comprehensive documentation of all studies of the evolutionary rates of proteins and DNA. Instead we focus on work that has been important in the historical development of ideas, or which illustrates particular issues.

The evolutionary events discussed in this book occurred over a 250 million year (Ma) period, during the Mesozoic and Cenozoic eras.

Frequent reference will be made in some sections of the book (particularly in chapters 7 to 10) to the various sub-eras, periods and epochs that make up these eras. The time-intervals that these represent may be unfamiliar to many readers. Table 1.1 provides a ready reference to help interpret the discussion in these sections.[6]

We hope that this book will provide a useful supplement to previous reviews of the molecular clock.[7-10] We mentioned at the beginning of this chapter that the molecular clock hypothesis remains controversial. In this book we present an account of this controversy as it has developed historically, and we show why different researchers have sometimes reached contrary conclusions from analysis of seemingly similar data. We find little convincing evidence that would lead to the rejection of a DNA clock in mammals, except in a few specific cases, but we do find evidence that leads to the rejection of at least some protein clocks. We are led to conclude that on current evidence the DNA clock hypothesis has *not* been rejected. We are, in fact, inclined to go further than this, and so we argue that there is now sufficient evidence to assert that in mammals nonrepetitive DNA evolves in a clock-like fashion.

This leads us to put forward a number of ideas that will be controversial. We suggest, for instance, that the radiation of placental mammals occurred long before the mass extinction at the end of the

Table 1.1 Geologic time scale for the last 250 million years (Ma)[6]

Era	Sub-era	Period	Epoch	Time Since Beginning (Ma)
Cenozoic	Quaternary	Pleistogene	Holocene	0.01
			Pleistocene	1.64
	Tertiary	Neogene	Pliocene	5.2
			Miocene	23.3
		Paleogene	Oligocene	35.4
			Eocene	56.5
			Paleocene	65.0
Mesozoic		Cretaceous	Gulf (Late)	97.5
			K_1 (Early)	145.6
		Jurassic	Malm (Late)	157.1
			Dogger (Middle)	178.0
			Lias (Early)	208.0
		Triassic	Tr_3 (Late)	235.0
			Tr_2 (Middle)	241.1
			Tr_1 (Early)	245.0

Cretaceous period that marked the end of the age of dinosaurs. We also argue that the divergence times of some primate taxa occurred substantially later than is currently thought and that this has important implications for human evolution. We question the long-standing assertion that there is sex difference in mutation rate associated with different rates of cell division in male and female germ lines. Similarly we question whether the differences in metabolic rate that exist among some mammalian species have any effect on the rate of mutation. We hope that by raising these issues our efforts will act as a stimulus for debate and further research in these areas.

References

1. Gould SJ. A clock of evolution. Nat Hist 1985; 4:13-25.
2. Kimura M. The neutral theory of molecular evolution. Cambridge: Cambridge University Press, 1983.
3. Nei M. Molecular evolutionary genetics. New York: Columbia University Press, 1987.
4. Gillespie J. The causes of molecular evolution. Oxford: Oxford University Press, 1991.
5. Hillis DM, Moritz C. Molecular systematics. Sunderland, Mass: Sinauer, 1990.
6. Harland WB, Armstrong RL, Cox AV, Craig LE, Smith AG, Smith DG. A geological time scale 1989. Cambridge: Cambridge University Press, 1989.
7. Wilson AC, Carlson SS, White TJ. Biochemical evolution. Ann Rev Biochem 1977; 46:573-639.
8. Thorpe JP. The molecular clock hypothesis: biochemical evolution, genetic differentiation and systematics. Ann Rev Ecol Syst 1982; 13:139-168.
9. Easteal S. A mammalian molecular clock? BioEssays 1992; 14:415-419.
10. Zuckerkandl E. On the molecular evolutionary clock. J Mol Evol 1987; 26:34-46.

CHAPTER 2

A Pattern Emerges: Early Comparison of Amino Acid Sequences

The impact of molecular biology on our understanding of biological evolution has been profound. The discipline of molecular evolution is now well established and it is hard to envisage a comprehensive evolutionary investigation that does not incorporate, if only indirectly, some component of molecular analysis. The fusion of molecular and evolutionary biology occurred during the 1960s and the molecular clock concept developed as one of the first important outcomes.

In the early part of this century the study of genetics and evolution, and of micro- and macro-evolutionary processes were quite separate. The connection had not been made between major evolutionary trends (the development of organ systems and the fate of species over geologic time)[1,2] and the occurrence and behavior of genetic variation within species.[3-7] The synthesis of these two fields in the 1940s[8-10] emerged in the absence of an understanding of DNA as the material basis for inheritance. However, the realization that there was a connection between small-scale genetic variation and major evolutionary events was an important precondition for the introduction of molecular biology into evolutionary studies. It meant that, in a general sense, when the structure of DNA was discovered and the molecular mechanisms of inheritance and gene expression started to be understood, their relevance to evolutionary biology was quickly appreciated.

Evolutionary theory, however, was faced with the challenge of reconciling the particulate nature of inheritance with the seemingly continuous nature of most heritable variation and of evolutionary processes. This challenge continues in both a philosophical and a practical sense. As Lewontin[11] put it, in the context of measuring variation within species "What we can measure is by definition uninteresting and what we are interested in is by definition unmeasurable." One response to this challenge has been to shift the focus of investigation from gross phenotypic variation to discrete molecular variation—to attend to the measurably uninteresting. This approach, an apparent avoidance of the real issues, has been extremely fruitful. Its scope falls short of a full understanding of the evolutionary process; it has, however, opened up productive new areas of research with important contributions being made particularly in some areas such as phylogeny.

The significance of the new discoveries in molecular biology for understanding evolution could be quickly perceived because of the earlier synthesis of genetics and evolutionary biology. It would be some time, however, before techniques including amino acid sequencing[12] and protein electrophoresis[13] had advanced sufficiently to make possible the empirical investigation of molecular variation either within or among species.

DISCOVERY OF THE MOLECULAR CLOCK

Early molecular evolutionary studies involved analysis of protein variation. Amino acid sequencing techniques were applied to homologous proteins in different species. The significance of the sequence variation that was observed was debated from both evolutionary and biochemical perspectives. Alfinsen,[14] for example, suggested that the invariant amino acid residues were the "functional, or at least the most significantly functional, parts" of proteins. He suggested that variable regions have little or no adaptive function. Margoliash and Smith,[15] while agreeing that constrained regions may be functionally important, argued that changes in amino acid composition also reflected selective pressures.

Protein sequence data also provided a basis for investigating phylogenies. Traditional phylogenetic analysis was based mainly on morphology. It is difficult to obtain quantitative estimates of morphological divergence, and interpretation of these kinds of traits is complicated by the possibility of convergent evolution. Sequence data appeared to have the potential to overcome both these problems. The realization that macromolecular sequences contain a quantifiable record of their evolutionary history gave rise to the science of molecular phylogenetics.[16] It also raised questions about rates of molecular evolution.

The idea of a molecular clock arose early in the analyses of molecular evolutionary rates. Zuckerkandl and Pauling[17] suggested that, if the degree of amino acid difference between hemoglobin chains was a function of time since separation, it would be possible to date the divergence times of different members of the globin gene family (e.g. β- and δ-globins). Based on their estimate of the number of amino acid substitutions between human and horse α-globins (18 amino acids) and an assumption that these two species diverged from each other between 100 and 160 Ma ago, they estimated that one amino acid substitution had occurred approximately every 14.5 Ma. This gave estimates of the divergence times of β- and δ-globin, β- and γ-globin, α- and β-globin and α- and γ-globin of 44, 260, 565, and 600 Ma respectively.

Zuckerkandl and Pauling were aware of the many assumptions involved in deriving these estimates. They pointed out that the observed number of amino acid differences between polypeptide chains is likely to be less than the number of amino acid substitutions that occurred during the evolutionary divergence of the chains. When more than one substitution occurs at the same amino acid site, only one of the substitutions can be detected when sequences are compared. They made an informal correction to account for this in their comparison of human and horse α-globins; they estimated 18 amino acid substitutions from observing 15 amino acid differences. Their analysis did not provide evidence of a molecular clock; it assumed one.

The discovery of the "molecular evolutionary clock" is usually attributed to Zuckerkandl and Pauling[18] who coined the phrase. However, Margoliash[19] had earlier compared the amino acid sequences of cytochrome c from a range of taxa including mammals, birds, fish and fungi (Table 2.1). In addition to demonstrating that this protein is remarkably conserved over this vast taxonomic range, he noted that the number of amino acid substitutions among vertebrate species is approximately proportional to the time since the species diverged. Furthermore, as seen in Table 2.1, at each taxonomic level the degree of sequence difference between a group of related taxa and an outgroup to them is approximately the same. Thus, there are approximately 11 differences between all of the mammal sequences and the bird sequence. Similarly there are approximately the same number of differences between the different mammal and bird sequences and the fish sequence, and between all the animal sequences and the yeast sequence.

Table 2.1. The comparative analysis of cytochrome c by Margoliash[19] showing the number of amino acid diferences between species

	horse	pig	rabbit	human	chicken	tuna	yeast
horse		3	-	12	12	19	44
pig			-	-	10	17	43
rabbit				-	11	19	45
human					14	21	43
chicken						18	43
tuna							48

Modified and reproduced with permission from E Margoliash, Proc Natl Acad Sci (USA) 1963; 50:672-679.

Margoliash concluded:

> It appears that the number of residue differences between the cytochrome *c* of any two species is mostly conditioned by the time elapsed since the lines of evolution leading to these two species originally diverged.

He was, in effect, proposing a molecular clock for cytochrome c, although he cautioned that more sequences would be required before rate constancy could be confirmed.

It is interesting to note the difference between the conclusion reached by Margoliash and that reached by Goodman[20,21] from his serological comparison of various proteins among primate species. Goodman contended that, although there was a general trend of increasing protein difference with phylogenetic divergence, the rate of change had decreased in the lineage leading to humans. He was later to expand on this theme in what became known as the "hominoid slowdown." Goodman's conclusion was largely based on assumed divergence times of gibbons from great apes and of humans from chimpanzees of 40 Ma ago and 30-25 Ma ago respectively. Although these dates were consistent with the paleontological interpretation of the time, they are almost certainly incorrect (see chapters 5 and 8). Margoliash's approach, on the other hand, involved comparing the sequence divergence of a number of related species (e.g. mammals) relative to a more distantly related or outgroup species (e.g. a fish). In this way his conclusion, unlike Goodman's, did not depend on assumptions about the divergence times of the species being compared. They were thus less prone to misinterpretation.

The term "molecular clock" was coined by Zuckerkandl and Pauling.[18] They compared α- and β-globin chains of cows, horses and humans (Table 2.2). It was evident from these comparisons that the number of amino acid substitutions increased with distance of relationship among species. Zuckerkandl and Pauling noted, in particular, that in comparisons of β-globins and of γ-globins relative to α-globins the number of differences is approximately the same in all the lineages. It appeared that the rate of amino acid substitution is approximately uniform both among lineages and between β- and γ-globins.

Table 2.2. Number of differences between mammalian globin chains presented by Zuckerkandl and Pauling.[18] The comparisons of β- and γ-globins relative to α-globins, which were most important in influencing their interpretation, are boxed

	horse α	cow α	human β	horse β	cow β	human γ	cow γ
human α	17	27	74	81	75	79	82
horse α		38	77	75	77	77	77
cow α			81	83	83	81	88
human β				26	27	39	32
horse β					35	43	33
cow β						45	28
human γ							40

Modified and reproduced with permission from E Zuckerkandl L. Pauling, in V. Bryson et al, eds. Evolving genes and proteins. New York: Academic Press, 1965:97-166.

Improving on their previous analysis, Zuckerkandl and Pauling introduced a formal method of estimating substitution rates from amino acid differences by assuming a Poisson model to describe the process of mutation. The same approach was put forward by Margoliash and Smith[15] and by Kimura.[22] Using Kimura's terminology, K_{aa}, the mean number of substitutions per amino acid site during an evolutionary period separating two polypeptides, can be estimated as:

$$K_{aa} = -ln(1-p_d), \tag{2.1}$$

where p_d is the fraction of amino acid sites that differ between the sequences. The rate of substitution per amino acid site per year (k_{aa}) is estimated from

$$k_{aa} = K_{aa}/(2T), \tag{2.2}$$

where T is the number of years since the divergence of the sequences from their most recent common ancestor. Applying this correction to their globin comparisons and assuming an 80 Ma divergence time between humans and artiodactyls they obtained estimates of the separation time between α- and β-/γ-globin genes of 365 Ma. This was substantially lower than their previous, provisional estimate of 565-600 Ma.

Analysis of globin genes was extended by Kimura,[22] who compared the α- and β-globins of various mammalian orders with the homologous genes in teleost fish (carp) and agnatha (lamprey). He noted that for both genes the rates in the different orders were approximately the same. He estimated rates of approximately 9×10^{-10} and 13×10^{-10} substitutions per amino acid site per year for α- and β-globin respectively by assuming a 500 Ma divergence of agnatha and mammals, a 375 Ma divergence time of mammals and fish, and an 80 Ma date for the radiation of placental mammals.

NATURAL SELECTION VERSUS GENETIC DRIFT

To many, a molecular clock was an unexpected result. Rates of morphological change, reflecting the process of adaptive evolution, were known to vary substantially among lineages.[23,24] If the raw material of adaptive evolution was mutational change in protein sequences, then proteins themselves would undergo adaptive evolution and this would be reflected in variation in the rate of evolution of their sequences.

Zuckerkandl and Pauling suggested that although adaptive evolution does involve changes in the function of proteins, these changes are the result of substitutions at only a small proportion of a protein's amino acid residues. Evolutionary divergence at these residues occurs at different rates in different lineages reflecting the process of adaptive evolution. However, molecular evolutionary change at most residues is independent of adaptive evolution. They suggested that the processes of morphological and molecular evolution were substantially independent of each other. The extent of functional change in a polypeptide chain was not necessarily proportional to the number of amino acid substitutions. Most substitutions have little or no effect on function; as Zuckerkandl and Pauling put it, they are "nearly neutral". This linked, in an informal way, the notion of neutrality of alleles, which had been investigated from a population genetics perspective for some time,[4,25] with evolution and the molecular clock.

Zuckerkandl and Pauling's[18] explanation for the molecular clock was, in effect, to propose a neutral theory of molecular evolution. This challenged prevailing orthodoxy in both population genetics and evolutionary biology. The new synthetic theory of evolution emphasized the role of natural selection. An explanation for the molecular clock involving mutations that were neutral to the action of natural selection was perceived as a major challenge by some of its proponents.[26] For many, evolution was too complex and varied a process to occur in a simple time-dependent fashion. Prominent among the opponents of rate constancy was Mayr[23] who argued that the rate of morphological evolution often varied following species divergence, especially in relation to the development of organ systems. Proteins involved in the function of such modified systems could be expected to evolve at a greater rate in the lineage in which the systems were developing. In a general sense he was arguing that the process of adaptive evolution involves changes in the function of proteins, and that such changes should lead to changes in the rates at which those proteins evolve.

Much later, Mayr[27] was still reluctant to accept evidence of a molecular clock on the basis that it conflicted with patterns of relatedness indicated by comparative morphology. He cited the small degree of molecular divergence between humans and their "distant relative" the chimpanzee, which is less than the divergence between different species of *Drosophila*. He maintained that even if protein evolution was in some cases approximately clock-like, "molecular clocks were governed by natural selection, not by rates of mutation". He even went as far as to propose that belief in the role of stochastic evolutionary processes was related to belief in Marxism.

If the evolutionary rate constancy seemed inconsistent with prevailing "pan-selectionist" views, it was not completely clear why it should be consistent with neutral evolution. Although intuitive arguments were put forward, the formal prediction of rate constancy under neutral evolution came later through the application of the mathematical theories of population genetics (chapter 3).

The problem of explaining a molecular clock in the absence of an appropriate theory is exemplified by Margoliash and Smith's[15] attempt to reconcile a molecular clock with pan-selectionism. They extended the earlier analysis of Margoliash[19] on cytochrome c, and concluded that both cytochrome c and the globin genes were evolving at a constant rate. They argued, however, that rate variation occurred over short periods of time in response to positive natural selection, but that over long periods of evolutionary time such effects would be averaged out across lineages giving the appearance of rate constancy.

It was apparent from the cytochrome c comparisons that some residues had evolved more rapidly than others. Margoliash[19] had concluded what was obvious: that if rapidly evolving regions of the molecule had a function then that function was compatible with a variety of primary structures. Margoliash and Smith[15] took the view that natural selection was the cause of the faster rate of evolution of some protein regions. They argued that "neutral mutations", since they were not preserved by natural selection, would be lost in the gene pool by dilution. If the function of a particular part of the molecule is equally well served by a variety of amino acid sequences, evolution would not have conserved any mutations in these regions. They suggest that the expectation is that functionally unimportant regions will be invariant in a homologous set of proteins. By implication, the variable regions of the cytochromes c are regions which have undergone unknown functional changes in different lineages.

The problem with these predictions, which ignore the role of random drift in evolution, were apparent to Margoliash and Smith[15] from their analysis of the distribution of variation within the cytochrome c molecule. There were several variable sites in the functionally important region between residues 11 and 33 and a completely invariant region between residues 70 and 80 with no known function. However, the carboxyl-terminal sequence was highly variable despite the fact that experimentally the cytochromes of different species could be exchanged without loss of efficiency of electron transport, and despite the fact that the last four residues of the protein could be removed without affecting the reaction rate. They concluded "it becomes difficult to maintain, at one and the same time, that the function of the carboxyl-terminal sequence of cytochrome c, if any, is compatible with an impressive array of radically different amino acids and that only those mutations which have a definite selective advantage will be conserved".

One pattern that emerged from these early comparative studies was that cytochrome c had evolved more slowly than had the globins. Analysis of differences in the evolutionary rates among proteins was taken further by Dickerson.[28] He investigated the pattern of sequence conservation and variation in cytochrome c in the context of the newly determined crystal structure and thereby explained some of the anomalies identified by Margoliash and Smith. He derived k_{aa} values of 4.5×10^{-9}, 8.6×10^{-10}, and 2.5×10^{-10} for fibrinopeptides, globins and cytochrome c respectively (Fig. 2.1). The globin rate is very similar to Kimura's estimate mentioned above. From these and other differences in rate Dickerson concluded that the evolutionary rates of proteins are dependent on the complexity of their interactions with other molecules.

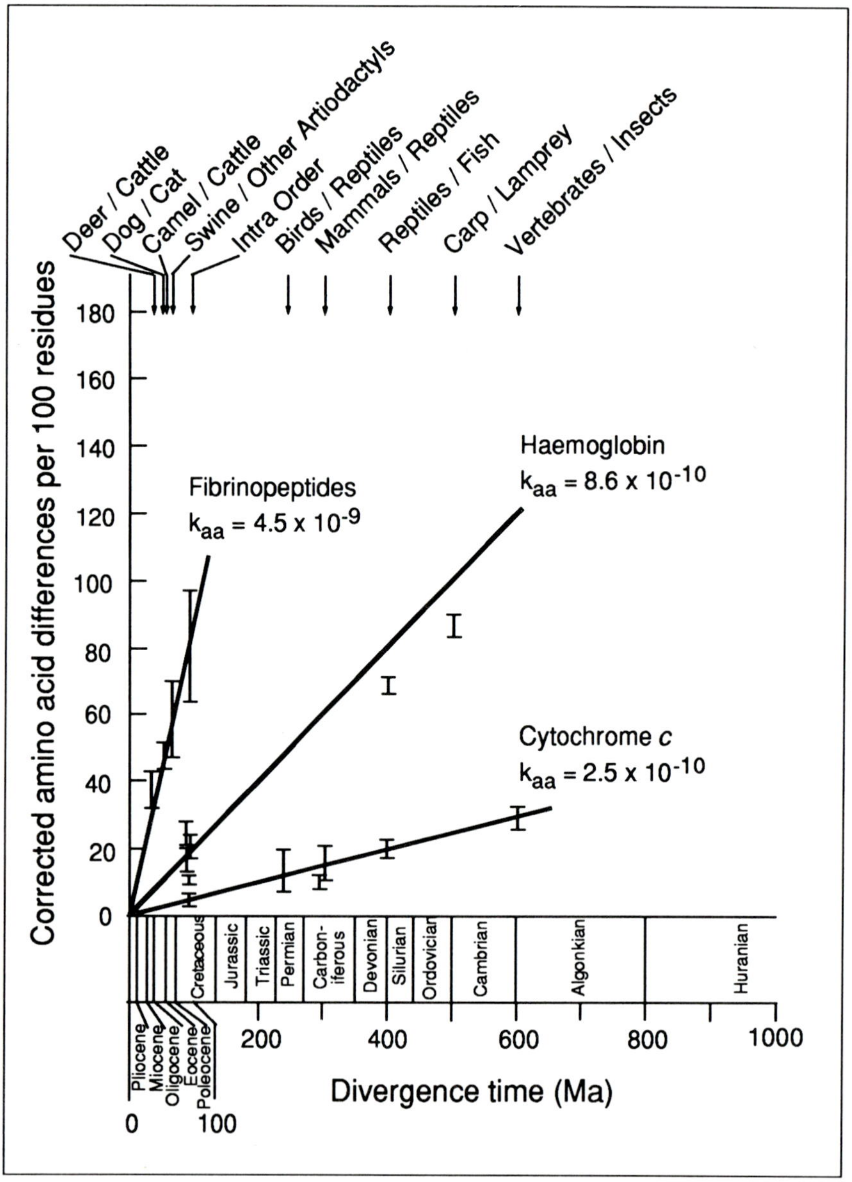

Fig. 2.1. Rates of evolution in the fibrinopeptides, hemoglobin and cytochrome c. The number of corrected amino acid differences is plotted as a function of divergence time. The evolutionary branching points for the different taxa are indicated by the arrows. The rate of amino acid substitution (k_{aa}) is given for each of the proteins. Modified and reprinted with permission from RE Dickerson, J Mol Evol 1971; 1:26-45.

In summary, the molecular clock hypothesis was perceived as an important challenge to the synthetic theory of evolution. This challenge led some to reject the hypothesis outright. It led others to attempt to reconcile the molecular clock with the pan-selectionism inherent in the synthetic theory. Most importantly it stimulated the suggestion that genetic drift rather than natural selection may play the dominant role in the evolution of proteins. The full implications of the molecular clock in this respect would become clear when Kimura considered the issue in the context of the mathematical theories of population genetics.

References

1. Gregory WK. Evolution emerging: a survey of changing patterns from primeval life to man. New York: Macmillan, 1951.
2. de Beer GR. The evolution of the metazoa. In: Huxley J, Hardy AC, Ford EB, eds. Evolution as a process. London: George Allen and Unwin, 1953:24-33.
3. Fisher RA. The genetical theory of natural selection. Oxford: Oxford University Press, 1930.
4. Wright S. On the roles of directed and random changes in gene frequency in the genetics of populations. Evolution 1948; 2:279-294.
5. Haldane JBS. The causes of evolution. New York; Longmans, Green, 1932.
6. Ford EB. Ecological Genetics. London: Chapman and Hall, 1975.
7. Dobzhansky T. Genetics and the origin of species. New York: Columbia University Press, 1937.
8. Huxley JS. Evolution, the modern synthesis. London: Allen & Unwin, 1942.
9. Simpson GG. The major features of evolution. New York: Columbia University Press, 1953.
10. Mayr E. Systematics and the origin of species. New York: Columbia University Press, 1942.
11. Lewontin RC. The genetic basis of evolutionary change. New York: Columbia University Press, 1974.
12. Sanger F. The free amino groups of insulin. Biochem J 1945; 39:507-515.
13. Markert CL. Biology of isozymes. In: Markert CL, ed. Isozymes I. Molecular structure. New York: Academic Press, 1975:1-9.
14. Alfinsen CB. The molecular basis of evolution. New York: Wiley, 1959.
15. Margoliash E, Smith EL. Structural and functional aspects of cytochrome *c* in relation to evolution. In: Bryson V, Vogel HJ, eds. Evolving genes and proteins. New York: Academic Press, 1965:221-242.
16. Crick FHC. On protein synthesis. Symp Soc Exptl Biol 1958; 12:138-163.
17. Zuckerkandl E, Pauling L. Molecular disease, evolution and genic heterogeneity. In: Kasha M, Pullman B, eds. Horizons in biochemistry. New York: Academic Press, 1962:189-225.
18. Zuckerkandl E, Pauling L. Evolutionary divergence and convergence in proteins. In: Bryson V, Vogel HJ, eds. Evolving genes and proteins. New York: Academic Press, 1965:97-166.

19. Margoliash E. Primary structure and evolution of cytochrome c. Proc Natl Acad Sci (USA) 1963; 50:672-679.
20. Goodman M. Evolution of the immunologic species specificity of human serum proteins. Hum Biol 1962; 34:104-150.
21. Goodman M. Serological analysis of the systematics of recent hominoids. Hum Biol 1963; 35:377-436.
22. Kimura M. The rate of molecular evolution considered from the standpoint of population genetics. Proc Natl Acad Sci (USA) 1969; 63:1181-1188.
23. Mayr E. Animal species and evolution. Cambridge: Harvard University Press, 1963.
24. Zuckerkandl E. On the molecular evolutionary clock. J Mol Evol 1987; 26:34-46.
25. Kimura K. Stochastic processes and distribution of gene frequencies under natural selection. Cold Spring Harbor Symp. Quant Biol 1955; 20:33-51.
26. Simpson GG. Organisms and molecules in evolution. Science 1964; 146:1535-1538.
27. Mayr E. The growth of biological thought: diversity, evolution and inheritance. Cambridge: Harvard University Press, 1982.
28. Dickerson RE. The structure of cytochrome c and the rates of molecular evolution. J Mol Evol 1971; 1:26-45.

CHAPTER 3

A Theory to Match: The Neutral Theory of Molecular Evolution Predicts a Molecular Clock

Uniformity in the rate of evolution of cytochrome c and of the globins contrasted with variation in the rate of morphological evolution. The resolution of this discrepancy suggested by Zuckerkandl and Pauling[1] was that evolutionary changes in gross phenotype could be accounted for by changes at only a small proportion of the amino acid sites in proteins. Different amino acids at the remaining sites were, in effect, neutral to the action of natural selection.

THE COST OF NATURAL SELECTION

Margoliash[2] and Zuckerkandl and Pauling[1] put forward the molecular clock hypothesis on the basis of similarities in evolutionary rate among different lineages determined by reference to an outgroup species (Tables 2.1 and 2.2). The hypothesis did not depend on the particular rate of substitution per year of the proteins being compared. Substitution rates were, however, estimated by reference to the fossil record. Kimura[3] estimated an evolutionary rate of approximately one substitution per 100 amino acid sites per 28 Ma by averaging the rates estimated for cytochrome c, hemoglobin, and triosephosphate dehydrogenase.[4] He then assumed: (1) A haploid genome size of 4,000 megabases (Mb)[5] (this is somewhat larger than the current estimate of 3,000 Mb for the size of the human genome[6]), and (2) that 20% of nucleotide replacements are synonymous, i.e., that one amino acid replacement corresponds to 1.2 base pair replacements (this was based on the further, obviously incorrect assumption that the entire genome consisted of coding sequences). On this basis he estimated that one nucleotide substitution had occurred approximately every two years in mammalian evolution.

The lack of precision of this estimate, based as it is on three proteins and questionable fossil-derived divergence times, and its inaccuracy following from the lack of knowledge of the extensive noncoding components of mammalian genomes, are obvious. However, the estimate is if anything too low, and the point Kimura was trying to make was that this is a very high rate indeed when compared with that predicted by Haldane[7] (one substitution every 300 generations).

Haldane had investigated the constraints on evolutionary rate imposed by the fact that natural selection involved differential reproduction and/or survival. There is, as he put it, a "cost", in terms of the survival of individuals in a population, of substituting one allelic form of a gene for another by natural selection (Kimura[3] referred to this as the substitutional load). Haldane showed that, except where selection was very intense, the overall cost (D) of substituting an allele depends on the initial frequency (p) of the selected allele but is independent of selection intensity. Thus, for semidominance,

$$D = -2 \ln p. \tag{3.1}$$

This was later modified by Kimura and Maruyama[8] thus:

$$D = -2 \ln p + 2, \tag{3.2}$$

to account for the effect of genetic drift in a finite population. Haldane estimated $D = 30$ by assuming $p = 5 \times 10^{-5}$. This is the frequency of a newly arisen mutant in a population of 10,000 individuals (containing 20,000 sets of chromosomes). He then arbitrarily assumed a selection intensity (I) of 0.1, selection intensity being the degree to which the fitness of the population is reduced relative to the optimum fitness which is equivalent to the proportion of the overall cost of substitution that is borne in each generation. The number of generations required for the substitution of a mutation was obtained as $D/I = 300$.

Kimura[3] pointed out that the load imposed by the substitution of one mutation every two years would be intolerably large. Later[9] he showed, by assuming that one generation is equal to one year, that one substitution every two years could only be achieved by a species in which only one of more than three million offspring survived. He concluded[3]:

> The very high rate of nucleotide substitution ... can only be reconciled with the limit set by the substitutional load by assuming that most mutations produced by nucleotide replacement are almost neutral in natural selection.

The details of Haldane's and Kimura's arguments about the rate of evolution depend on their assumptions about the size of populations and the intensity of selection. The load is less in smaller populations and fixation can occur more rapidly when selection intensity is greater. The argument also depends on an assumption of multiplicative fitness among loci. This assumption was questioned by Maynard Smith[10] and Sved.[11] They proposed threshold or truncation models of selection in which individuals that have advantageous alleles at more than some threshold number of loci survive irrespective of differences in the products of their fitness values at individual loci. These models (which Kimura[9] criticizes as unrealistic) allow much faster rates of substitution.

Irrespective of its ultimate validity, Haldane's model provided the theoretical framework for the analysis of molecular evolutionary rates that led to Kimura's proposition of the neutral theory. Kimura did not explicitly predict a molecular clock from theoretical population genetics. The prediction was, however, implicit in his reasoning; a fact not missed by King and Jukes[12] in their paper subtitled "most evolutionary change in proteins may be due to neutral mutations and genetic drift". They credit Kimura[3] with pointing out that "the *rate of random fixation* of neutral mutations in evolution (per species per generation) is equal to the *rate of occurrence* of neutral mutations (per gamete per generation)."

POPULATION GENETICS AND THE MOLECULAR CLOCK

The predicted association between mutation rates and fixation or substitution rates was based on Kimura's earlier theoretical work. In much of the early development of population genetics theory, the possibility of random changes in allele frequency (genetic drift) was not given serious consideration. Genetic drift results from the equivalent of sampling error in the selection of gametes during reproduction in each generation. The smaller the number of gametes that are involved in the production of a new generation (the sample size), the greater the random fluctuation in allele frequency from generation to generation. Such fluctuations can result in a newly arisen mutation eventually reaching complete fixation, particularly in small populations. However, Fisher[13] in particular considered that most natural populations would be so large that genetic drift would be of little importance in shaping the dynamics of allele frequency change.

The possible importance of genetic drift was recognized by Wright.[14] Wright also recognized that the effective size of a population (N_e), from a genetic point of view, may not be the same as the actual number of individuals (N) in the population. This will be the case if mating is not random, if the numbers of males and females differ, if generations are overlapping, or if the population is subdivided geographically. It is the effective size of the population that is important in determining the extent of genetic drift. It has now been shown that in many species that $N_e/N << 1$ and that even in numerous widespread species N_e may be small (e.g. ref. 15, 16).

An important development in the theoretical treatment of genetic drift was the use of diffusion equations.[17] Using this approach, Kimura[18,19] showed that in the case of semidominance (that is when both alleles in a heterozygote have equal dominance) the probability of fixation in the population of an allele with initial frequency p is given by

$$u(p) = (1 - e^{-4N_e sp})/(1 - e^{-4N_e s}), \qquad (3.3)$$

where s is the selection coefficient of the allele. The frequency of a newly arisen mutation will be $1/(2N)$, since initially it will occur on only one chromosome in one individual in the population. When s is small the probability of fixation of a new mutation is given by:

$$u = (4N_e s)/2N(1 - e^{-s}). \qquad (3.4)$$

When $N_e s << 1$, in other words for neutral alleles, this equation is reduced to:

$$u = 1/(2N). \qquad (3.5)$$

Kimura[3] argued from this that "new alleles may be produced at the same rate per individual as they are substituted in the population in evolution". At the time he did not make the reason for this explicit. It was made clearer in later publications.[9,20]

The rate of substitution of new mutations (k) depends on the rate at which they arise in the population (which is the product of the mutation rate per gamete (v) and the number of gametes or chromosomes in the population ($2N$)), and the probability of their becoming fixed (u). Thus

$$k = 2Nvu. \qquad (3.6)$$

For neutral mutations, equation (3.5) can be substituted in (3.6) to give:

$$k = v. \qquad (3.7)$$

Thus, for Kimura neutral mutations explained both the rapid rate and the constancy of rate of molecular evolution. The rate constancy prediction came from stochastic population genetics theory which also predicts a lack of rate constancy for mutations affected by natural selection. This is shown by substituting (3.3) in (3.6) to give

$$k \approx 4N_e sv. \qquad (3.8)$$

When mutants are substituted in populations by positive Darwinian selection the substitution rate depends on the size of the population and on the selection coefficient. Only in the unexpected case that $N_e s$ is constant will substitution rate also be constant.

This theoretical treatment of the issues confirmed the earlier intuitive arguments of Zuckerkandl and Pauling.[1] The molecular clock predicted by neutral theory is not expected to result from the substitution of mutations by natural selection. It thus provides a testable prediction of neutral theory. The prediction, however, depends on the uniformity of mutation rates. Mutation rates might be expected to vary for a number of reasons. These include variation in exposure to mutagens, changes in the fidelity or efficiency of DNA repair and replication enzymes, differences in the rate of cell replication in germ-lines, and differences in metabolic rate among species. The value of the molecular clock as a test of neutral theory is somewhat limited when substitution rate variation can be attributed to mutation rate variation. Mutation rate variation resulting particularly from DNA-replication rate differences or generation time differences are, in fact, frequently invoked to explain observed differences in rate.

This issue will be discussed extensively later in this book. The question of whether DNA replication rate affects mutation rate depends on the cause of the mutations that are substituted in the course of evolution. If these mutations are the result of errors during DNA replication then clearly replication rate will have an effect. If, however, they are primarily the result of DNA-repair errors then a replication rate effect is not expected.

In this context, transgenic mice have recently been developed which have in their genomes reporter genes that allow the direct measurement of mutation rates *in vivo*. In the first study using this method to investigate the effect of mitotic rate on mutation rate *no effect was observed.*[21] Mutation rates in the rapidly dividing cells of the spleen are the same as those in the slowly dividing cells of the kidney and brain. This important result for somatic cells will need to be confirmed in germ cells before any definite evolutionary conclusions can be drawn from it. It does, however, provide direct evidence that *the rate of cell division does not affect the rate of mutation.*

King and Jukes[12] suggest that Kimura proposed that for neutral mutations the rate of substitution per generation would equal the rate of mutation per generation. In fact, Kimura did not specify any unit of time and his proposition can equally well be represented by the statement: the rate of random fixation of neutral mutations in evolution (per species *per year*) is equal to the rate of occurrence of neutral mutations (per gamete *per year*). There is some suggestion of the importance of generation time in Kimura's paper in the sense that Haldane's arguments are phrased in terms of generation time. Kimura[3] also discussed the compatibility of substitution rates with mutation rates based on biophysical considerations, and in doing so assumed that mutations result from DNA replication errors. However, Kimura[22] did state that:

> The remarkable constancy per year [of amino acid substitutions in hemoglobins] is most easily understood by assuming that in diverse vertebrate lines the rate of production of neutral mutations per individual per year is constant.

Kimura and Ohta[20] also made it clear that the rate constancy prediction does not depend on mutation rates being generation time-dependent. They stated that:

> The uniformity of the rate of mutant substitution per year for a given protein may be explained by assuming constancy of neutral mutation rate per year over diverse lines.

The neutral theory prediction about substitution rate depends on the assumption made about mutation rate. If mutation rate is constant per generation, neutral theory predicts that substitution rate will be constant per generation. If mutation rate is constant per year, neutral theory predicts that substitution rate will be constant per year. It is clear, however, that in the formulation and early development of neutral theory, Kimura and Ohta were influenced by the results of the comparative amino acid sequence studies. From these it was concluded that substitution rates (and hence, under neutral theory, mutation rates) were constant per year. The perception that generation time is important in determining molecular evolutionary rates would come later through the analysis of the evolutionary rates of DNA, discussed in the next chapter.

BIOCHEMISTRY, MOLECULAR BIOLOGY AND THE NEUTRAL THEORY

Kimura[9,23] has pointed out that, in addition to the arguments developed from theoretical population genetics, neutral evolution is consistent with the biochemical properties of amino acids, the nature of the genetic code, and the structure of the eukaryote genome. Neutral theory was, in essence, proposed on these grounds both by Zuckerkandl and Pauling[1] and by King and Jukes.[12]

King and Jukes pointed out, for example, that substituted amino acids in cytochrome c often have similar chemical and physical properties, arguing that such substitutions may have little effect on the function of the molecule. In particular they drew attention to the common occurrence of substitutions involving the very similar amino acids valine, leucine and isoleucine (Table 3.1).

They also pointed out that many of the mutant hemoglobins in humans that appear harmless are characterized by amino acid differences on the outside of the molecule and that these differences are the same as differences that exist between humans and other species (Table 3.2). In this way they made a connection between variation *among* species and polymorphism *within* species. This connection was made much more strongly by Kimura[3] and Kimura and Ohta,[20] the latter entitling their paper "Protein Polymorphism as a Phase of Molecular Evolution".

Table 3.1. Amino acid substitutions in cytochrome c suggested by King and Jukes[12] to be neutral because they are conservative

	Amino acid site in cytochrome c						
Species	**17**	**19**	**43**	**65**	**66**	**93**	**103**
Neurospora	Leu	Lys	Leu	Ile	Thr	Leu	Ile
Saccharomyces	Leu	Lys	Ile	Val	Leu	Leu	Ile
Candida	Leu	Lys	Ile	Val	Glu	Leu	Val
Wheat	Ile	Lys	Leu	Val	Glu	Leu	Ile
Moth	Ile	Val	Phe	Ile	Thr	Leu	Ile
Horse	Ile	Val	Leu	Ile	Thr	Ile	Ile
Other	Val	Ile	Val		Ile/Val		

Reproduced with permission from JL King, T H Jakes, Science 1969; 164:788-798.

Table 3.2. α- and β-globin amino acid residues that are both variant forms in humans and normal forms in other species

	Amino acid site in α- globin				Amino acid site in β-globin			
	22	**57**	**68**	**68**	**16**	**69**	**87**	**95**
Human (normal)	Gly	Gly	Asn	Asn	Gly	Gly	Thr	Lys
Human (mutant)	Asp	Asp	Lys	Asp	Asp	Asp	Lys	Glu
Other species aa	Asp	Asp	Lys	Asp	Asp	Asp	Lys	Glu
Other species	Carp	Orangutan	Rabbit Sheep	Carp	Horse	Cow	Pig Rabbit	Pig

Reproduced with permission from JL King, T H Jakes, Science 1969; 164:788-798.

NEUTRAL THEORY AND GENETIC VARIATION

The connection between the evolution of molecular variation between species and the genetics of molecular variation within species was important in understanding the significance of the extensive molecular variation identified in populations of humans[24] and *Drosophila.*[25] This variation was detected by histochemical staining of specific protein variants separated on the basis of electrostatic charge using starch-gel electrophoresis. In both species it was found that an average individual was heterozygous at a substantial proportion of loci. Subsequent investigations have shown this to be true of most species.[26]

Lewontin[27] has discussed the alternative views about genetic variation that prevailed before this extensive molecular variation was discovered. These views were referred to as the "classical" and the "balance" views by Dobzhansky.[28] The classical view derives from Muller[29] and his early contribution to classical genetics. Briefly, the view is that the genome is essentially monomorphic except for the presence of recessive deleterious mutations. Advantageous mutations arise occasionally but are rapidly fixed in populations of the species and do not contribute substantially to variation. The balance view,[28] which arose largely from study of chromosomal and morphological variation in natural populations by Dobzhansky, Ford[30] and others, was that species are highly polymorphic with variation being maintained by balancing natural selection.

The balance view would seem to have been vindicated by the molecular data except that Kimura and Crow[31] had earlier shown that, because of the segregational load it would impose, balancing selection could only account for polymorphism at a small number of loci. They were careful to point out, however, that the occurrence of extensive variation could not be ruled out if much of the variation were nearly neutral.

The neutral theory explanation of protein polymorphism was vigorously opposed, particularly by students of Dobzhansky.[27,32-34] The segregational load argument was criticized in much the same way as was Kimura's substitutional load argument mentioned above. Truncation selection models were put forward,[34-36] which are discussed in more detail by Gillespie.[37] There followed what became known as the neutralist-selectionist debate. Detailed consideration of this debate is beyond the scope of this book. Discussion of it can be found elsewhere.[9,27,38]

What we would emphasize here is the point made by Kimura and Ohta,[20] that molecular evolutionary rates and levels of molecular polymorphism are interdependent. It is worth noting that many recent studies designed to test neutral theory are based on analysis of patterns of molecular variation both within and among species (e.g. ref. 39).

The issue of generation-time effects on mutation rate is important in this context. Kimura and Crow[31] showed that the probability that an individual is heterozygous in a population is:

$$H_e = 4N_e v_g /(4N_e v_g + 1), \tag{3.9}$$

where v_g is the mutation rate *per generation*. Lewontin[27] used the implicit assumption that mutation rate is generation–time dependent to argue from equation (3.9) that the pattern of genetic variability among species is incompatible with neutral theory, stating that:

> Since there is no reason to suppose that mutation rate has been specifically adjusted in evolution to be the reciprocal of population size for higher organisms, we are required to believe that higher organisms including man, mouse, Drosophila and the horseshoe crab all have population sizes within a factor of 4 of each other. ... The patent absurdity of such a proposition is strong evidence against a neutralist explanation of observed heterozygosity. (p 208-210)

If, however, mutation rate were constant per year, there might in fact be a reciprocal relationship between N_e and v_g, with a much narrower range of variability predicted among species. As Kimura and Ohta[20] put it:

> We note that if the mutation rate *u* is constant per year, then the product *Neu* should be less variable among different organisms than its components *Ne* and *u*, because the species with short generation time tends to have small body size and attain a large population number.

This issue was investigated by Nei and Graur[40] who found no significant correlation of generation time with population size. There may, however, have been substantial errors in their estimates of population size, and they took no account of the difference between effective and actual population size. The difficulty of obtaining reliable estimates of effective population size may preclude any accurate assessment of its relationship to generation time in different species. The point to be made, however, is that this relationship will have an important bearing on expected heterozygosity if mutation rate is constant per year. It will have no bearing on the expectation if mutation rate is constant per generation.

References

1. Zuckerkandl E, Pauling L. Evolutionary divergence and convergence in proteins. In: Bryson V, Vogel HJ, eds. Evolving genes and proteins. New York: Academic Press, 1965:97-166.
2. Margoliash E. Primary structure and evolution of cytochrome c. Proc Natl Acad Sci (USA) 1963; 50:672-679.
3. Kimura M. Evolutionary rate at the molecular level. Nature 1968; 217:624-626.
4. Kaplan NO. Evolution of dehydrogenases. In: Bryson V, Vogel HJ, eds. Evolving genes and proteins. New York: Academic Press, 1965:243-277.
5. Muller HJ. Evolution by mutation. Bull Amer Math Soc 1958; 64:137-160.
6. Fields C, Adams MD, White O, Venter JC. How many genes in the human genome? Nature Genet 1994; 7:345-346.
7. Haldane JBS. The cost of natural selection. J Genet 1957; 55:511-524.
8. Kimura M, Maruyama T. The substitutional load in a finite population. Heredity 1969; 24:101-114.
9. Kimura M. The neutral theory of molecular evolution. Cambridge: Cambridge University Press, 1983.
10. Maynard Smith J. "Haldane's dilemma" and the rate of evolution. Nature 1968; 219:1114-1116.
11. Sved JA. Possible rates of gene substitution in evolution. Am Nat 1968; 102:283-293.

12. King JL, Jukes TH. Non-Darwinian evolution. Science 1969; 164:788-798.
13. Fisher RA. The genetical theory of natural selection. Oxford: Oxford University Press, 1930.
14. Wright S. On the roles of directed and random changes in gene frequency in the genetics of populations. Evolution 1948; 2:279-294.
15. Easteal S. The ecological genetics on introduced populations of the giant toad *Bufo marinus*. II. Effective population size. Genetics 1985; 110:107-122.
16. Easteal S, Floyd RB. The ecological genetics of introduced populations of the giant toad, *Bufo marinus* (Amphibia: Anura): dispersal and neighbourhood size. Biol J Linn Soc 1986; 27:17-45.
17. Kimura K. Stochastic processes and distribution of gene frequencies under natural selection. Cold Spring Harbor Symp Quant Biol 1955; 20:33-51.
18. Kimura M. Some problems of stochastic processes in genetics. Ann Math Stat 1957; 28:882-901.
19. Kimura M. On the probability of fixation of mutant genes in a population. Genetics 1962; 47:713-719.
20. Kimura M, Ohta T. Protein polymorphism as a phase of molecular evolution. Nature 1971; 229:467-469.
21. Lee AT, DeSimone C, Cerami A, Bucala R. Comparative analysis of DNA mutations in *lacI* transgenic mice with age. FASEB J 1994; 8:545-550.
22. Kimura M. The rate of molecular evolution considered from the standpoint of population genetics. Proc Natl Acad Sci (USA) 1969; 63:1181-1188.
23. Kimura M. Molecular evolutionary clock and the neutral theory. J Mol Evol 1987; 26:24-33.
24. Harris H. Enzyme polymorphism in man. Proc Roy Soc London Ser B 1966; 164:298-310.
25. Lewontin RC, Hubby JL. A molecular approach to the study of genic heterozygosity in natural populations. II. Amount of variation and degree of heterozygosity in natural populations of *Drosophila pseudoobscura*. Genetics 1966; 54:595-609.
26. Nevo E. Genetic variation in natural populations: patterns and theory. Theor Pop Biol 1978; 13:121-177.
27. Lewontin RC. The genetic basis of evolutionary change. New York: Columbia University Press, 1974.
28. Dobzhansky T. A review of some fundamental concepts and problems of population genetics. Cold Spring Harbor Symp Quant Biol 1955; 20:1-15.
29. Muller HJ. Our load of mutations. Am J Hum Genet 1994; 2:111-176.
30. Ford EB. Ecological genetics. London: Chapman and Hall, 1975.
31. Kimura M, Crow JF. The number of alleles that can be maintained in a finite population. Genetics 1964; 49:725-738.
32. Powell JR. Isozymes and non-Darwinian evolution: a re-evaluation. In: Markert CL, ed. Isozymes IV genetics and evolution. New York:

Academic Press, 1975:9-26.
33. Ayala FJ. Darwinian versus non-Darwinian evolution in natural populations of *Drosophila.* Proc 6th Berkeley Symp Math Stat Prob 1972; 5:211-236.
34. Richmond R.C. Non-Darwinian evolution: a critique. Nature 1975; 225:1025-1028.
35. King JL. Continuously distributed factors affecting fitness. Genetics 1967; 49:561-576.
36. Sved JA, Reed TE, Bodmer WF. The number of balanced polymorphisms that can be maintained in a natural population. Genetics 1967; 55:469-481.
37. Gillespie J. The causes of molecular evolution. Oxford: Oxford University Press, 1991.
38. Nei M. Molecular evolutionary genetics. New York: Columbia University Press, 1987.
39. Hudson RR, Kreitman M, Aguadé M. A test of neutral molecular evolution based on nucleotide data. Genetics 1987; 116:153-159.
40. Nei M, Graur D. Extent of protein polymorphism and the neutral mutation theory. Evol Biol 1984; 17:73-118.

CHAPTER 4

Generation Gap: The Theory Is Modified when DNA Is Compared

The early comparative studies that led to the development of the molecular clock hypothesis involved amino acid sequences. These provided information about only a minute component of the genome and the information was indirect. Furthermore, amino acid sequencing techniques were expensive, labor intensive and slow; few laboratories could undertake the large-scale sequencing projects that comparative analysis required. There was a need for more accessible methods for directly investigating the evolution of genomes. Nucleotide sequencing techniques were not developed until much later,[1,2] and until the 1960s genomes were studied almost entirely by microscopy. Comparisons were made of the gross morphology of chromosomes arrested at metaphase. Although differential staining of chromatin components provided some resolution of the underlying DNA-protein structure, the level of understanding that could be achieved from this approach was crude. Nevertheless, comparative cytogenetics flourished for a time and was particularly important in the development of ideas about speciation.[3] Comparative cytogenetics was not, however, able to provide any quantitative measure of species divergence.

During the 1960s, the quantitative comparison of genomes became possible through the development of methods for analyzing the reassociation kinetics of heteroduplex DNA, i.e. DNA duplexes consisting of complementary strands from different species.[4-6] This approach held great promise because with relative technical ease the entire genomes of different species could potentially be compared.

These early DNA-DNA hybridization studies indicated a component of eukaryote genomes (approximately 20%) that is completely conserved among even distantly related species.[6] This was inconsistent with results from prokaryotes and suggested that a substantial component of the genome is not involved in coding for proteins. The reassociation kinetics of homoduplexes (i.e. DNA duplexes consisting of complementary strands from the same individual) demonstrated that the genomes of higher eukaryotes consist of different components distinguishable by the degree of repetition of their sequences.[7,8] The organization of the eukaryote genome was revealed as a complex of highly-repetitive and middle-repetitive DNA elements interspersed among the (presumably) protein-coding single-copy sequences. The distribution and abundance of the repeated DNAs was found to differ significantly even between closely related species.[7] This meant that the repeated sequence components could not provide quantitative measures of evolutionary distance. Nonrepeated DNA sequences were thought to occur only once per haploid genome and, when comparing species, to be homologous in the sense that they are derived from a common ancestor.[7,8] They thus became the focus of evolutionary studies.

In this chapter we discuss the early work in which DNA-DNA hybridization methods were used to study rates of genome evolution. The conclusions drawn from these studies were different from those reached from the comparative analysis of protein sequences. The work of Margoliash[9] and Zuckerkandl and Pauling[10] suggested that the rate of amino acid replacement was approximately constant for any given protein independent of lineage, although the rate varied among proteins depending on functional constraint (chapter 2). In contrast, DNA-DNA hybridization studies suggested that evolutionary rate varied substantially among lineages.[8,11,12] Species with short generation times appeared to have higher rates of evolutionary change in their DNA than those with long generation times. This discrepancy gave support to the suggestion that mutation rates are generation-time dependent.[13] To put the arguments into context, and because DNA-DNA hybridization has largely passed into disuse, some explanation of the technique is required.

DNA-DNA HYBRIDIZATION METHODS

In the study of genome structure total DNA is sheared to an average length of approximately 500 bases. It is then denatured by heating to 95°C and quickly cooled to 65°C to allow reassociation. The degree of repetition of a sequence determines its kinetics of reassociation. Sequences repeated thousands or millions of times per genome reassociate more quickly than those present in low copy number. The amount of reassociation after set time intervals can be measured by the chromatographic separation of double and single-stranded DNAs. The number and distribution of repetitive sequences is informative from the viewpoint of genomic structure. For quantitative evolutionary analyses, however, repetitive sequences represent noise. They are removed by chromotographic separation following repeated rounds of preincubation of the DNA solution.

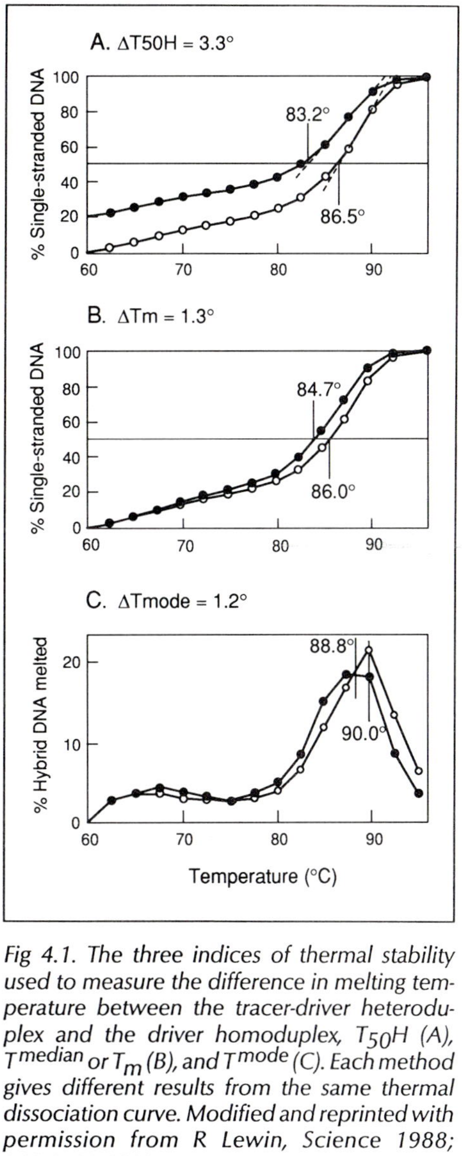

Fig 4.1. The three indices of thermal stability used to measure the difference in melting temperature between the tracer-driver heteroduplex and the driver homoduplex, $T_{50}H$ (A), T^{median} or T_m (B), and T^{mode} (C). Each method gives different results from the same thermal dissociation curve. Modified and reprinted with permission from R Lewin, Science 1988; 241:1598-1600.

In molecular evolutionary studies, single-copy DNA from a reference species is made radioactive and then mixed with the unlabeled DNA from a species to be tested. The radioactive "tracer" DNA and cold "driver" DNA are mixed in a ratio of approximately 1:10,000 so that effectively all tracer DNA strands reassociate with driver DNA strands after denaturation and cooling. The resultant duplex molecules are bound to a column. The temperature of the column is raised by defined small increments and the single-stranded DNA is eluted and assayed for radioactivity. The cumulative radioactivity released from the column is plotted against temperature. The temperature at which half the strands have dissociated is termed the melting temperature (T_m). Sequence divergence in homologous regions of DNA between tracer and driver species results in mispairing in the heteroduplex. The greater the sequence divergence the lower the melting temperature of the heteroduplex DNA. The difference in melting temperatures between tracer-driver heteroduplex and the driver homoduplex can be used as the measure of sequence divergence. Three indices of thermal stability are used, $T_{50}H$, T^{median} and T^{mode} (Fig. 4.1).

$T_{50}H$ is the temperature at which 50% of the total tracer DNA is dissociated from the heteroduplex. T_m measures the same parameter but after correction has been made for the difference in hybridization between heteroduplexes and homoduplexes. T^{mode} is the temperature at which the largest percentage of hybridized tracer DNA has dissociated.

In all three methods, the depression in melting temperature between the homoduplexes and the heteroduplexes is called the "T" value and is used to calculate sequence divergence. $T_{50}H$ is considered by some to be an inaccurate measure of thermal stability when the rate of evolution varies among different regions of the genome.[14-16]

The advantage claimed for DNA-DNA hybridization studies over sequence comparisons was that they examined the genetic affinities of the whole of the nonrepetitive DNA component, as opposed to a minute sample of the genome as represented by one or two protein sequences.[8,12] This point has been made more recently by Diamond.[17]

EARLY EXPERIMENTAL RESULTS

Laird, McConaughy and McCarthy[11] were the first to examine the extent of genetic divergence between single-copy mammalian DNAs using heterologous DNA/DNA duplex formation. They performed two experiments. In the first, labeled cow DNA was reacted with the unlabeled DNAs of cow (control), sheep, pig and opossum (*Didelphis virginianus*). In the second, mouse was used as the tracer DNA and reacted to the driver DNAs of rat, hamster and guinea pig, as well as with mouse as a control.

The results of the analyses are presented in Table 4.1 and Figure 4.2. In the comparisons using cow DNA as a tracer, the degree of sequence divergence increases in a linear fashion with the proposed time of separation of the species. As expected the opossum is the most divergent from cow in terms of both years and number of nucleotide substitutions. The comparison of the rodent DNAs, however, shows a larger degree of divergence between mouse and the other rodents than that seen in the artiodactyl comparison (including opossum).

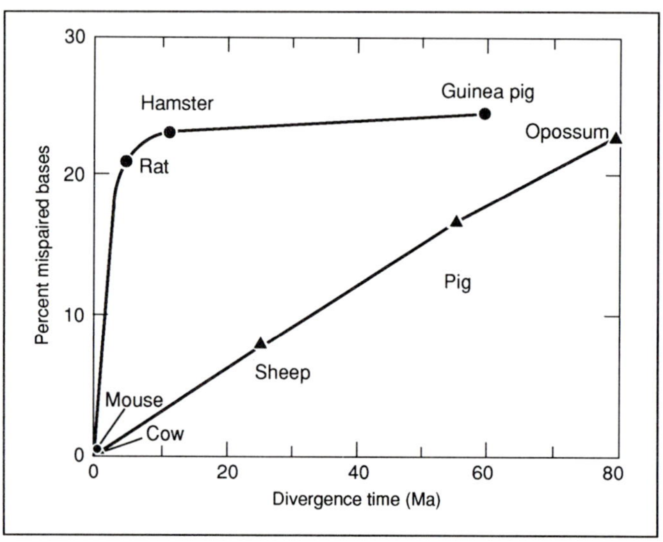

Fig. 4.2 The proposed rates of nucleotide divergence within the rodent lineage (relative to the mouse) and within the artiodactyls (relative to the cow) based on the percentage of mispaired bases in DNA-DNA hybridization studies. Modified and reprinted with permission from CD Laird et al, Nature 1969; 224:149-154.

Table 4.1. Laird et al's[11] estimates of ΔT_m and of the percent of mispaired bases between mammalian species

Species compared	ΔT_m	% base mispairing
sheep-cow	5	7.5
pig-cow	11.5	16.8
opossum-cow	15	22.5
rat-mouse	14	21
hamster-mouse	15	22.5
guinea pig-mouse	16	24

Modified and reproduced with permission from CD Laird et al, Nature 1969; 224-149-154.

From these analyses the authors concluded that the rate of fixation of nucleotide substitutions in the rodent lineage is approximately 10-fold higher than that seen in the artiodactyl lineage. This difference in rate was negated if generation time rather than absolute time was used to compare sequence divergence, with two to three generations per year assumed for rodents and 0.3 to 0.5 generations per year assumed for artiodactyls.

This analysis was done when methods of studying DNA reassociation kinetics were relatively new and some comment is needed on specific technical aspects of the work, as distinct from the general limitations of DNA-DNA hybridization methods.

First, Laird et al[11] noted that in rodents the relationship between genome difference and species divergence is not linear. The genome of guinea pigs appears to have diverged from that of mice less than have the genomes of rats and hamsters in relation to the respective divergence times of these species. The effect of multiple substitutions at the same nucleotide site were not taken into account in deriving the genome divergence estimates and the possibility that the limits of applicability of the technique had been reached in making the more distant comparisons was not considered. It is known that the sensitivity of the technique begins to be reduced as sequences reach 15% divergence.[14] When levels of divergence exceed 20%, sequences will not hybridize.[16] This is evidenced as an increase in nonreassociated tracer DNA. The estimates of percent mispaired bases in most of Laird et al's comparisons exceed 15%, and in many cases the proportion of tracer DNA that reassociates is low. The values for comparisons of pig-cow, opossum-cow, rat-mouse, hamster-mouse and guinea pig-mouse are respectively 50%, 26%, 38%, 20%, and 6%.

Second, the mouse and cow tracer DNAs were treated differently and this may have caused some differences between the results obtained for rodents and artiodactyls. Of some concern, then, is the pooling of results from the two different experiments to allow comparison and the lack of reciprocal hybridizations to test validity of the results. The conclusion that generation time is important depends very much on the result of the mouse-rat comparison, for which, as Brownell[18] has shown, reciprocal reactions are very important because of the large difference in genome size. Having said this, however, it should be pointed out that the result obtained by Laird et al[11] for the mouse-rat comparison is not substantially different from those obtained by other workers.[18,19]

Third, the 1200 bp DNA fragments generated for reassociation were longer than the usual 500 bp.[7] This difference would alter the kinetics of reassociation. It would increase the rate of reassociation but decrease the extent of reassociation between heteroduplexes. The temperature increments of the thermal gradient are also large, over 5°C intervals, which introduces uncertainty in the estimates of melting temperature.

We have discussed Laird et al's results in some detail because of their significance as the first attempt to quantify the differences between mammalian genomes and because they had a strong influence on thinking about rates of molecular evolution. For all the reasons mentioned above, the accuracy of their sequence divergence estimates is questionable. Also questionable are their assumptions about the times of divergence of the species they were comparing.

We will discuss issues about the estimation of species divergence times in more detail later. However, the point is worth making here that in molecular evolutionary rate studies there has been a tendency for molecular biologists and statistical geneticists to interpret the palaeontological literature in a rather naïve way. Secondary or tertiary sources are often cited and too little attention is given to the tentative nature of many interpretations and extrapolations.

It was pointed out in the last chapter that the difference between Goodman's studies, from which he concluded that evolutionary rate varies, and the studies of Margoliash and of Zuckerkandl and Pauling, which led to the conclusion of rate constancy, was that the former made assumptions about species divergence times while the latter did not. The discrepancy is probably explained by Goodman's assumptions being wrong. Laird et al's conclusions also depend on the validity of species divergence times estimated from the fossil record. They cite Simpson[20] for rodent divergence times and Matthew[21] for artiodactyl divergence times.

Simpson's article is a general theoretical discussion about systematics which draws on a number of examples including Wood's[22] phylogeny of the rodents. In discussing the Muridae (mice and rats), Wood makes no comment on the divergence of mice and rats from each other. He does, however, draw attention to dispute about the age of the family Muridae. He believes that it is recent, but cites Ellerman[23] who argued that members of the family were previously arboreal and have a long unrecorded history. The Muridae are related to the Cricetidae (voles, lemmings and hamsters). Wood gives the earliest Cricetidae as

Lower Oligocene. The general view is that the Muridae diverged later than this, but this may not be the case and there is in fact no real evidence of when the divergence actually occurred. Hence, Laird et al's estimates of 5 Ma and 12 Ma ago for the dates of the rat-mouse and murid-hamster divergences cannot be derived from their cited source.

The assumption about the divergence of the Myomorpha (murids and hamsters) from the Caviomorpha (guinea pigs) is worse. With respect to the Myomorpha, Wood states "The origin of this suborder can only be guessed at." He also discusses the debate about the origins of Caviomorpha, concluding that it is unresolved (he does have his own ideas, which, in light of what is now known about continental drift, are almost certainly wrong). Laird et al's assumed 60 Ma divergence of Myomorphs and Caviomorphs is based on Simpson's use of dotted lines to represent what Wood concedes is guesswork. Simpson's dotted lines join horizontally in the early Eocene, but this is not a reflection of any paleontological evidence. It has more to do with the fact that the beginning of the Eocene is also the border of Simpson's diagram. We will discuss more recent evidence relating to rodent divergence times in chapter 7.

There are also problems with Laird et al's[11] estimates of artiodactyl divergence times. They cite Matthew[21] for cow/sheep and cow/pig divergence times of 25 Ma and 55 Ma respectively. Matthew's phylogenetic diagram, from which these estimates are derived, showed bovids (labeled "antelopes, cattle etc.", which includes sheep) as first appearing in the early Miocene (say 20-23 Ma ago) when they separated from prong-horns (Antilocapridae). Any split between sheep and cattle should logically be more recent than this. Thus according to Matthew, the source of Laird et al's estimate, cows and sheep must have diverged much later than the 25 Ma they assumed.

Laird et al's estimate of 55 Ma for the time of the split between the pigs and their relatives (bunodonts or Suiformes) and selenodonts (camels plus ruminants) is more soundly based. Matthew placed this divergence in the early to mid Eocene (45-55 Ma ago) and more recent reviews of the artiodactyl fossil record generally support Matthew.[21] According to Romer[24] and Carroll,[25] definite suiforms and selenodonts first appear in the mid-Eocene, and the first ruminants appear in the late Eocene. Gentry and Hooker's[26] interpretation is that the major artiodactyl clades had already separated by the early Eocene (50-55 Ma) and Webb and Taylor[27] imply that pecorans were distinct by the late Eocene and that therefore ruminants appeared earlier.

Finally, Laird et al give a divergence time of placental mammals from the marsupial opossum of 80 Ma without citing any source for this date. As we will see later, it is almost certainly too late, maybe by as much as a factor of two.

The theory that DNA substitutions accumulate at a rate dependent on generations rather than years was supported by Kohne.[8] He analyzed the divergence of the single-copy sequences of primates. Tracer DNAs from a human and an African green (or vervet) monkey were each reassociated with driver DNAs from, in ascending taxonomic distance from humans, a human, a chimpanzee, a gibbon, a rhesus monkey, an African green monkey, a capuchin (a species of New World monkey), and a galago (bush baby). From these, he calculated the degree of divergence on each branch of the primate lineage. Using divergence times derived from interpretation of the primate fossil record, they then estimated the rate of nucleotide substitution on each of the branches (Fig. 4.3). Kohne concluded that both the extent of thermal reassociation and the thermal stability of the heteroduplexes agree with theories of primate relationships based on palaeontological and comparative anatomical evidence.

Kohne[8] also claimed that the rate of sequence evolution has decreased in the lineage leading to humans. However, evidence for this is hard to find. From Figure 4.3 it appears that the rate of substitution is particularly high in segment E. However, the nucleotide difference estimates involving galago are, by the author's own admission, unreliable, and the apportioning of nucleotide change on branches E and L is, in the absence of an outgroup, arbitrary (he appears to recognize the latter point in that he refers to the E, L, D, and K branch estimates as "estimates"). The other segment in which the rate of change appears high is D. Galagos are a reliable outgroup to monkeys and apes; however, since the distance estimates involving galago are unreli-

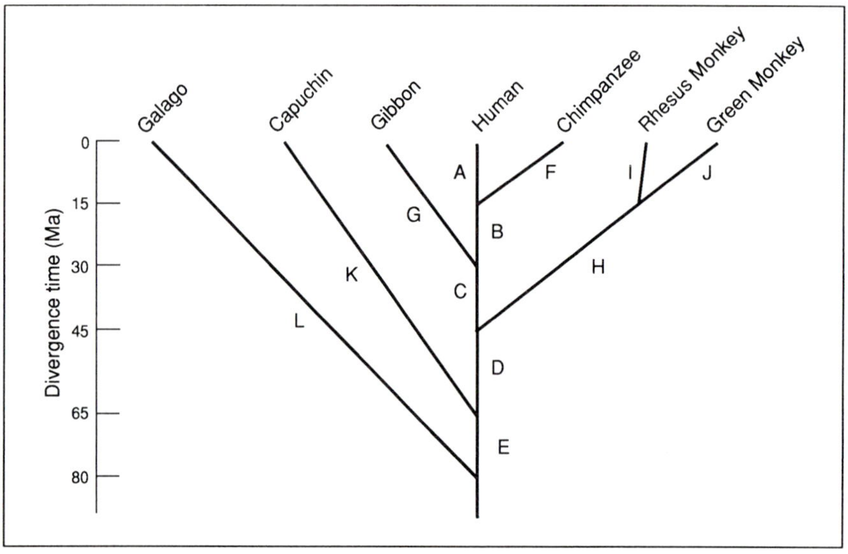

Fig. 4.3. The distribution of nucleotide changes on the branches of primate phylogenetic tree according to Kohne.[8] Modified and reprinted with permission from DE Kohne, Quart Rev Biophys 1970; 3:327-375.

able, the apportioning of nucleotide change to segments D and K is again arbitrary. The remaining estimates all appear to be approximately the same, despite errors in the assumed divergence times, some of which are probably substantial. The rest of Kohne's argument for a generation-time effect is based on a comparison of his results with those obtained for rodents by Laird et al,[11] the inadequacies of which we have already discussed.

It is worth noting that had Kohne used the relative approach to comparing rates adopted by Margoliash and by Zuckerkandl and Pauling he might have arrived at a different conclusion. His data show that humans, chimpanzees and gibbons have all diverged from African green monkeys to about the same degree (10.5, 10.8 and 10.5% nucleotide difference respectively). They also show that the human-capuchin divergence (17.4%) is about the same as the African Green monkey divergence (18.5%). This difference translates into a 20% difference in rate between the human and African green monkey lineages since their divergence from each other. Given the limitations of the data this small difference is probably of no significance.

Kohne's[8] interpretation of his own and Laird et al's[11] data is summarized in Figure 4.4A. Rodents appear to have experienced a faster rate of nucleotide substitution than artiodactyls, which in turn appear

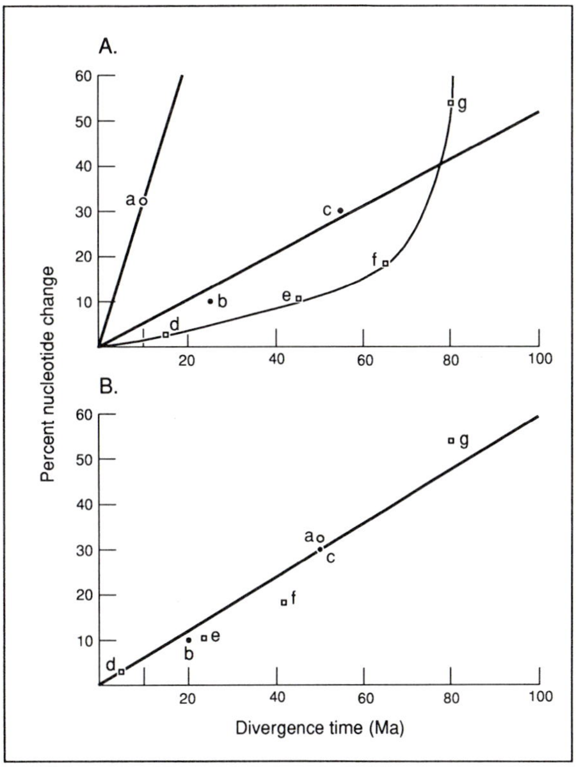

Fig. 4.4. Percent nucleotide change as a function of time since divergence in three orders of placental mammals. A. Based on divergence times assumed by Kohne.[8] B. Based on divergence times discussed in chapters 7 and 8.

to have evolved more rapidly than primates. However, when the same nucleotide substitution data are considered in the light of our discussion of fossil-derived divergence times (chapters 7 and 8), there is no apparent evolutionary rate difference among the three orders.

Kohne argued that the rate of accumulation of mutations is constant if generation time is taken into account. More specifically, he concluded that the rate of nucleotide substitution correlates with the number of germ cell divisions from zygote to zygote. DNA replication was thus invoked as the predominant mutational mechanism in molecular evolution.[8,12] This conclusion, despite the inadequacy of the data on which it was based and of the interpretation of those data had a profound effect on the development of molecular evolutionary studies.

THE NEARLY NEUTRAL THEORY OF MOLECULAR EVOLUTION

The discrepancy between the conclusions of the comparative amino acid sequence studies and the DNA-DNA hybridization studies presented a challenge to the newly formulated neutral theory. If mutation rates depended on generation time, as the DNA-DNA hybridization studies were interpreted as suggesting, then why were proteins evolving at rates that seemingly did not depend on generation time? Neutral theory predicted that if mutation rate varied with generation time then so would substitution rate. Kimura[28] and Kimura and Ohta[29] had, after all, suggested that the constant rate of hemoglobin evolution was best explained by assuming that the rate of mutation per individual per year was constant.

If it was difficult to reconcile the presumed difference in the pattern of protein and DNA evolution with neutral theory, it was just as difficult to reconcile it with positive Darwinian selection. In the previous chapter, we saw from equation (3.8) that when mutations are substituted by positive selection, substitution rate is proportional to the product of effective population size and selection coefficient (N_es). Generation time-dependent mutation rates would imply that

$$k = N_e sv/g, \tag{4.1}$$

If k is constant, the product of effective population size and selection coefficient would have to vary in a way that exactly compensated for any variation in mutation rate among lineages due to differences in generation time. Such a relationship is highly improbable. Even in the extraordinary circumstance that selection coefficients remained constant, the relationship implies that species with shorter generation times have smaller effective population sizes. If anything, as Kimura and Ohta[29] pointed out, the opposite is expected.

An alternative solution was put forward by Ohta and Kimura[30] and Ohta.[31] Kimura[32] had suggested that "nearly neutral" mutations, i.e., those for which $|2N_e s| << 1$ would behave in a very similar way to strictly neutral mutations. Ohta and Kimura[30] clarified the concept of "nearly neutral" mutations making the point that there is a continuum, and not a clear distinction, between neutrality and non-neutrality (Fig. 4.5). For slightly deleterious mutations, if mutation rate is generation time dependent, then:

$$k \propto v/(N_e sg) \qquad (4.2)$$

(compare this with equation (4.1)).

Ohta[31] stated that "We know as an empirical fact, that there is a negative correlation between population size and generation time", although she provided no source or evidence for this fact. She argued that, since this is the case, for slightly deleterious mutations (which would all have very low, and therefore similar, selection coefficients) the effects of generation time and effective population size would cancel each other out. The result would be that mutations would be substituted at a rate that was constant per year, despite a mutation rate that depended on generation time. We have already mentioned Nei and Graur's[33] failure to find a correlation between population size and generation time. Chao and Carr,[34] however, have shown that a correlation does exist between $\log(N)$ and $\log(g)$, and have suggested that this is the more appropriate relationship in the context of Ohta's model.

Models of the evolution of slightly deleterious mutations have been extended[35-40] and are discussed in more detail elsewhere.[41-43] They were born of the need to reconcile the discrepant, and quite possibly erroneous conclusions of early imperfect studies of the rates of DNA and protein evolution. The discrepancies between these studies are probably more a reflection of inadequate methodology than of any real difference in the patterns of DNA and protein evolution. The importance of slightly deleterious mutations in molecular evolution does not,

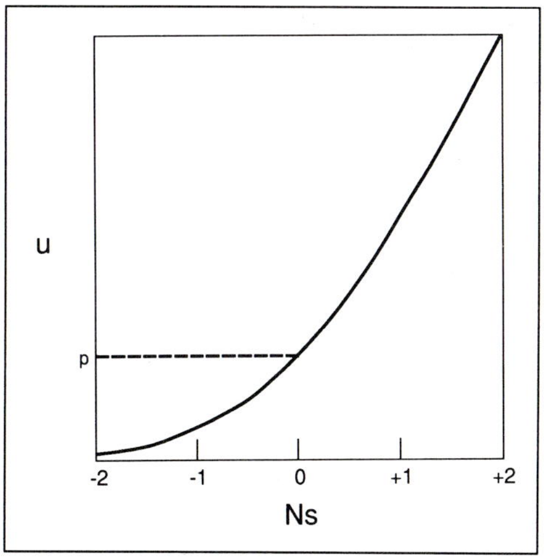

Fig. 4.5. Fixation probability (u) of mutant allele with initial frequency (p) as a function of the product of population size (N) and selection coefficient (s). For neutral mutations u = p. Modified and reprinted with permission from T Ohta et al, J Mol Evol 1971; 1:18-25.

however, depend on the validity of the conclusions that they were invoked to explain. If mutation rate were independent of generation time *and* if a substantial proportion of substituted mutations were slightly deleterious, a quite different pattern of molecular evolution would be observed. Rates of noncoding DNA evolution would be constant per year (reflecting the rate constancy of mutations), and rates of amino acid sequence evolution would vary in relation to variation in N_e/g. In chapter 6 we will discuss more recent evidence in relation to this issue.

References

1. Sanger F, Nicklen S, Coulson AR. DNA sequencing with chain-terminating inhibitors. Proc Natl Acad Sci (USA) 1977; 74:5463-5467.
2. Maxam AM, Gilbert W. A new method for sequencing DNA. Proc Natl Acad Sci (USA) 1977; 74:560-564.
3. White MJD. Animal cytology and evolution. Cambridge: Cambridge University Press, 1977.
4. McCarthy BJ, Bolton ET. An approach to the measure of genetic relatedness among organisms. Proc Natl Acad Sci (USA) 1963; 50:156-164.
5. Marmur J, Falkow S, Mandel M. New approaches to bacterial taxonomy. Ann Rev Microbiol 1963; 17:329-372.
6. Hoyer BH, McCarthy BJ, Bolton ET. A molecular approach to the systematics of higher organisms. Science 1964; 144:959-967.
7. Britten RJ, Kohne DE. Repeated sequences in DNA. Science 1968; 161:529-540.
8. Kohne DE. Evolution of higher-organism DNA. Quart Rev Biophys 1970; 3:327-375.
9. Margoliash E. Primary structure and evolution of cytochrome *c*. Proc Natl Acad Sci (USA) 1963; 50:672-679.
10. Zuckerkandl E, Pauling L. Evolutionary divergence and convergence in proteins. In: Bryson V, Vogel HJ, eds. Evolving genes and proteins. New York: Academic Press, 1965:97-166.
11. Laird CD, McConaughy BL, McCarthy BJ. Rate of fixation of nucleotide substitutions in evolution. Nature 1969; 224:149-154.
12. Kohne DE, Chiscon JA, Hoyer BH. Evolution of primate DNA sequences. J Hum Evol 1972; 1:627-644.
13. Goodman M. Evolution of the immunologic species-specificity of human serum proteins. Hum Biol 1962; 34:104-150.
14. Houde P. Critical evaluation of DNA hybridization studies in avian systematics. Auk 1987; 104:17-32.
15. Cracraft J. DNA hybridization and avian phylogenetics. Evol Biol 1987; 21:47-96.
16. Springer M, Krajewski C. DNA hybridization in animal taxonomy: a critique. Quart Rev Biol 1989; 64:291-318.
17. Diamond JM. Taxonomy by nucleotides. Nature 1983; 305:17-18.
18. Brownell E. DNA/DNA hybridization studies of muroid rodents: symme-

try and rates of molecular evolution. Evolution 1983; 37:1034-1051.
19. Chevret P, Denys C, Jaeger J-J, Michaux J, Catzeflis FM. Molecular evidence that the spiny mouse (*Acomys*) is more closely related to gerbils (Gerbillinae) than to true mice (Murinae). Proc Natl Acad Sci USA 1993; 90:3433-3436.
20. Simpson GG. The nature and origin of supraspecific taxa. In: Anonymous, ed. Genetics and Twentieth Century Darwinism. New York: Cold Spring Harbor, 1959:255-271.
21. Matthew WD. A phylogenetic chart of the artiodactyla. J Mammal 1934; 15:207-209.
22. Wood AE. A revised classification of the rodents. J Mammal 1955; 36:165-185.
23. Ellerman JR. The families and genera of living rodents. London: British Museum (Nat. Hist.), 1949.
24. Romer AS. Vertebrate paleontology. 1966; Chicago: University of Chicago Press.
25. Carroll RL. Vertebrate paleontology and evolution. New York: W.H. Freeman & Co., 1988.
26. Gentry AW, Hooker JJ. The phylogeny of the Artiodactyla. In: Benton MJ, ed. The phylogeny and classification of the tetrapods, Volume 2: Mammals. Oxford: Clarendon Press, 1988:235-271.
27. Webb S, Taylor BE. The phylogeny of hornless ruminants and a description of *Archeomeryx*. Bull Amer Mus Nat Hist 1980; 167:117-158.
28. Kimura M. The rate of molecular evolution considered from the standpoint of population genetics. Proc Natl Acad Sci (USA) 1969; 63:1181-1188.
29. Kimura M, Ohta T. Protein polymorphism as a phase of molecular evolution. Nature 1971; 229:467-469.
30. Ohta T, Kimura M. On the constancy of the evolutionary rate of cistrons. J Mol Evol 1971; 1:18-25.
31. Ohta T. Evolutionary rate of cistrons and DNA divergence. J Mol Evol 1972; 1:150-157.
32. Kimura M. Evolutionary rate at the molecular level. Nature 1968; 217:624-626.
33. Nei M, Graur D. Extent of protein polymorphism and the neutral mutation theory. Evol Biol 1984; 17:73-118.
34. Chao L, Carr DE. The molecular clock and the relationship between population size and generation time. Evolution 1993; 47:688-690.
35. Ohta T. Population size and rate of evolution. J Mol Evol 1972; 1:305-314.
36. Ohta T. Role of very slightly deleterious mutations in molecular evolution and polymorphism. Theor Pop Biol 1976; 10:254-275.
37. Ohta T. Very slightly deleterious mutations and the molecular clock. J Mol Evol 1987; 26:1-6.
38. Ohta T. Theoretical study of near neutrality. II. Effect of subdivided population structure with local extinction and recolonization. Genetics

1992; 130:917-923.

39. Ohta T, Tachida H. Theoretical study of near neutrality. I. Heterozygosity and rate of mutant substitution. Genetics 1990; 126:219-229.
40. Kimura M. A model of effectively neutral mutations in which selective constraint is incorporated. Proc Natl Acad Sci (USA) 1979; 76:3440-3444.
41. Gillespie J. The causes of molecular evolution. Oxford: Oxford University Press, 1991.
42. Kimura M. The neutral theory of molecular evolution. Cambridge: Cambridge University Press, 1983.
43. Ohta T. The nearly neutral theory of molecular evolution. Ann Rev Ecol Syst 1992; 23:263-286.

CHAPTER 5

Monkey Business: Molecular Evolution in Primates

The investigation of primate evolution has always engendered controversy, reflecting, no doubt, the primate nature of the investigators. Molecular evolution is no exception; some of the earliest studies of the molecular clock focused on the primates, and a substantial body of primate molecular data suitable for comparative analysis is now available. The debate about primate molecular clocks is ongoing and in this chapter we discuss its history.

The existence of extensive background knowledge of general primate biology and evolution is a very good reason for focusing on primates to study molecular evolutionary issues. Particularly important is the degree of confidence that can be placed in current theories about the order in which other major primate lineages have branched from the human lineage. There is now no serious disagreement that first to branch were the strepsirhines (lemurs, lorises and galagos), followed by tarsiers, New World monkeys, then Old World monkeys. Among the apes, gibbons diverged first, followed by the orangutan and then the African apes. Molecular data have been important in confirming phylogenetic inferences drawn from comparative morphology and paleontology. They have also resolved some uncertainties including the position of tarsiers,[1-3] and the closer affinity of chimpanzees to humans than to gorillas.[4-17]

THE RELATIVE RATE TEST

This well established branching order makes possible the use of the relative approach to analyzing molecular evolutionary rates, pioneered by Margoliash[18] and first applied to primates by Sarich.[19] The principle behind this approach is illustrated in Figure 5.1

Three species (or sequences) are usually compared. The two more closely related species are the subject of the rate test (X and Y in

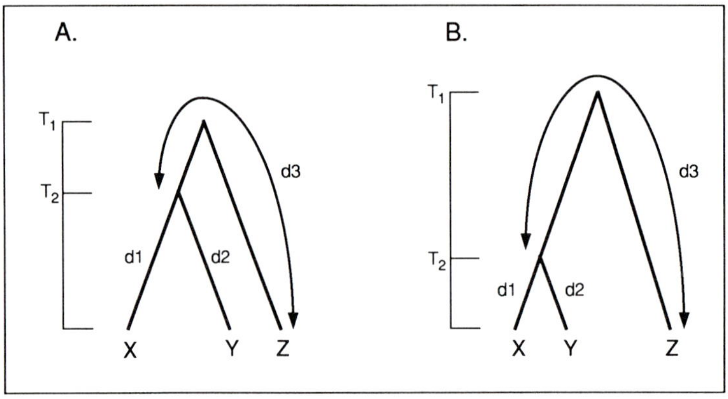

Fig. 5.1. Relative and absolute rate methods of testing the molecular clock for three species, X, Y, and Z which diverged at times T_1 and T_2, and which differ by molecular distances d1+d2, d1+d3, and d2+d3. The relative approach involves no assumption about T1 or T2. A molecular clock is rejected if d1+d3 ≠ d2+d3. The absolute approach, in which a molecular clock is rejected if, say, (d1+d2)/T2 ≠ (d1+d3)/T1, can give very different results depending on the assumed values of T1 and T2 (this is illustrated by comparing phylogeny A with phylogeny B). Modified and reprinted with permission from Easteal, BioEssays 1992; 14:415-419.

Fig. 5.1A), the third, more distantly related species (Z) is used as a reference for comparison of the first two. The extent of nucleotide change along the individual branches of the tree relating the three species cannot be determined directly from the sequence comparison, but cumulative changes in the form of distances can. Thus the distance between X and Z is d1+d3, and the distance between Y and Z is d2+d3. The relative test of evolutionary rates involves the comparison of d1+d3 with d2+d3. A difference between them implies a difference between d1 and d2, since d3 is common to both. The important point is that the test does not require knowledge of the specific divergence times, T_1 and T_2, of the compared species. The validity of a test result would not depend on whether divergence times are as shown in phylogeny A or phylogeny B, although the power of this approach to detect rate differences is diminished if widely divergent reference species are used. The important assumption that underlies this approach is that the reference species, Z, is the most distantly related of the three. Because the primate branching order is so well known, appropriate reference species are readily identified.

The relative rate approach produces unambiguous results when appropriate reference species can be identified. This contrasts with the ease with which erroneous conclusions about evolutionary rates can be reached when these depend on the interpretation of the fossil record to derive species' divergence times. This point, which was made in the previous chapter, is illustrated by again considering Figure 5.1. Evolutionary rates are estimated by assuming divergence times T_1 and T_2 in phylogeny A, and a molecular clock is rejected if, for example, $(d1+d2)/T_2 \neq (d1+d3)/T_1$. The molecular clock may have been falsely rejected if the real divergence times T_1 and T_2 were as shown in phylogeny B.

We repeat this point here because, despite seeming to be obvious, it has not been appreciated by many investigators of molecular evolutionary rates. In primates the results of early relative and fossil-record based studies were very different. The resolution of this difference resulted in a drastic reinterpretation of the fossil record.

EARLY ANALYSIS OF GLOBIN SEQUENCES

Analysis of molecular evolution in primates goes back to the work of Goodman[20,21] who measured immunological differences among primates for a range of proteins. To interpret these differences in terms of evolutionary rates he used fossil evidence to estimate divergence times. The result was his proposal that the rate of evolution had slowed down in the lineage leading to humans, which he referred to as the "hominoid slowdown".

He suggested a number of reasons for the slowdown. One, an increase in generation time in humans relative to other primates, still has some support and will be discussed in more detail later. Another, which he advocated more strongly at the time, was the "placentation hypothesis": that there is selection against new mutations in placental mammals due to feto-maternal incompatibility. The extent of this purifying selection is greater in species with more developed placenta and longer gestation periods, such as humans. This idea was an extension to other proteins of the feto-maternal incompatibility of different rhesus blood group types, which had recently been identified as a cause of disease in human neonates. However, rhesus incompatibility is unusual, and although incompatibility occurs occasionally with respect to other red cell blood groups (mainly ABO) it is much less severe. Similar conditions for other proteins are not currently a recognized clinical problem, and hence Goodman's placentation hypothesis does not seem to be of any general significance.

Goodman and colleagues, through their study of globins, have been persistent advocates of the importance of natural selection in molecular evolution. They have related apparent accelerations in evolutionary rate to major advances in physiology, and decelerations in rate to the effect of stabilizing selection on well adapted molecules.[22-25]

Their arguments reveal an implicit belief in evolutionary advancement towards a superior biological state in "higher" primates, culminating in *Homo sapiens*.[25] An assumption is made that primates have better "homeostasis" than other placental mammals, although homeostasis is never clearly defined and no evidence of its greater development in primates is ever provided. Goodman et al,[22] for example, in discussing the advantage of proline at certain positions in the hemoglobin molecule suggested that "the greater incidence of proline at such positions in human haemoglobin than in ancestral primitive haemoglobins [meaning those of other species] may be attributed to positive selection bringing about further improvements". Goodman et al[24] continued in this vein by asserting that their alleged globin gene slowdown in hominoids points to "selection preserving perfected haemoglobin molecules" and that it indicates "selection in the pre-hominines of haemoglobin molecules with finely tuned, perfected adaptations".

The obvious anthropocentrism exhibited by Goodman and his colleagues does not necessarily invalidate their conclusions about evolutionary rates. However, the conclusions are invalidated by the methods of analysis used in deriving them. The divergence times assumed for both species and duplicated genes, the methods of correcting for multiple substitutions, and the methods of assigning sequence change to particular branches of phylogenetic trees have all been shown to be inappropriate.[26-29] This criticism is reviewed by Kimura[29] and we will not repeat it here. Suffice it to say that there is now no basis for continuing acceptance of their interpretation and their analysis serves only to illustrate the way in which inappropriate assumptions and inadequate methods of analysis can lead to erroneous conclusions. The practice of estimating the length of branches in phylogenetic trees from poor fossil evidence and the use of such estimates to argue for a "hominoid slowdown" has, however, persisted. This approach to the analysis of nucleotide sequence data is discussed below.

IMMUNOLOGICAL DISTANCES

Sarich and Wilson[30] extended Goodman's early work on the immunological differences between homologous proteins in different primate species using the micro-complement fixation (MC'F) technique.[31] Antibody was prepared against serum albumin from four primate species (human, chimpanzee, an Old World monkey and a New World monkey). These were individually reacted with a variety of heterologous (different species) albumins and with homologous (same species) albumin. Immunological cross-reactivity between two species was measured by a quantity *R*, the ratio of the amount of antibody needed to titrate heterologous antigen to the amount needed to titrate homologous antigen.

Sarich and Wilson[30] confirmed the result that there is an association between the presumed degree of species' relatedness and the extent of immunological cross-reactivity or distance. Thus, they showed that *R* values for comparisons of humans with chimpanzees, Old World monkeys and New World monkeys were approximately 1.1, 2.3 and 5.7 respectively. They then investigated the relative rates of albumin evolution[19] by comparing the degree of immunological cross-reactivity within several groups of primate species relative to a reference outgroup species. In this way they demonstrated that the rate of albumin evolution was approximately the same within simians (relative to tarsiers and galagos), within the catarrhines (relative to New World monkeys), within the New World monkeys (relative to catarrhines), within the Old World monkeys (relative to apes), and within the apes (relative to Old World monkeys). Their results for apes, catarrhines, and simians are shown in Table 5.1. By looking down the columns of the table it can be seen that the immunological distances between the different members of a group of taxa and their common reference species are approximately the same. Sarich and Wilson concluded:

> During the approximately 45 million years that have elapsed since apes, man, Old World monkeys and New World monkeys last shared a common ancestor, the various lineages leading to the modern species have experienced similar amounts of albumin evolution.

This result contrasted with Goodman's[21,22] conclusion that the rate of molecular evolution had slowed down in the human lineage. Since Goodman's conclusions depended on the interpretation of the primate fossil record current at the time, it followed that this interpretation might not be correct.

Sarich and Wilson[32] went on to consider this issue in more detail. They first derived the relationship between degree of immunological

Table 5.1. Immunological distances between the albumins of different pairs of primate taxa

	Reference species		
Compared species	**OWM**	**NWM**	**Tarsier**
Apes			
Human	2.09	4.2	10.0
Chimpanzee	2.17	3.9	
Gorilla	2.13		
Orangutan	2.17		
Gibbon	2.07		
Siamang	2.05		
Old World monkeys			
Maccaca		3.8	9.8
Cercopithecus		3.6	
Presbytis		4.2	
New World monkeys			
Cebus			9.0

Modified and reproduced with permission from V Sarich, Proc Natl Acad Sci (USA)1967; 58:142-148.

crossreactivity (R) and divergence time (T) from previous analysis of nonmammalian vertebrates as:

$$\log R = kT, \tag{5.1}$$

where k is a constant (a more theoretical approach to this relationship was developed later[34]). They estimated $k = 0.012$, by taking their observed value of $R = 2.3$ for the comparison of apes and Old World monkeys, and assuming that $T = 30$ Ma for these two taxa (which is probably too early by a substantial margin, see chapters 8 and 10). Applying this to their observed value of $R = 1.13$ for the divergence of humans from the other African apes they obtained $T = 5$ Ma. Similarly, they obtained estimates of 8 Ma and 10 Ma for the divergences of African apes from the orangutan and from gibbons respectively.

Previously much earlier divergence times had been presumed, and initially these dates were widely rejected by physical anthropologists, although, as Sarich and Wilson pointed out at the time, their dates provided solutions to a number of long-standing problems in the interpretation of the morphological relationships of fossil and extant apes. Since then a number of fossil apes have been discovered and others have been reinterpreted and there is now widespread acceptance that

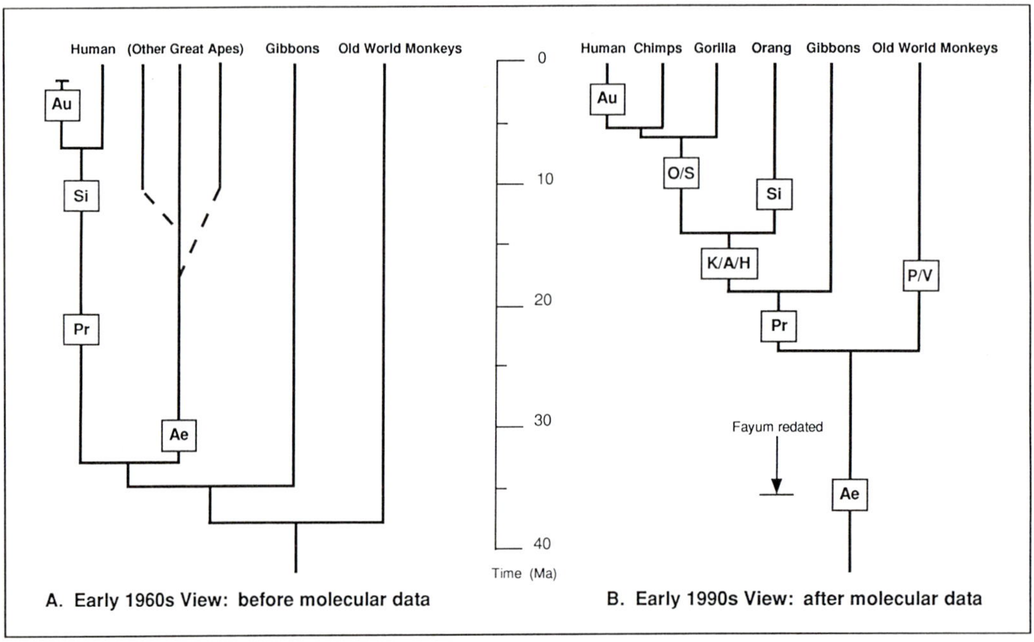

Fig. 5.2. Prevailing views about the pattern of catarrhine evolution. A. In the early 1960s, before the impact of molecular data. B. In the early 1990s, after the acceptance of the implications of molecular data, and the discovery or reinterpretation of numerous fossils. Au: Australopithecus, *Si:* Sivapithecus, *Pr:* Proconsul, *Ae:* Aegyptopithecus, *O:* Ouranopithecus, *S: Samburu mandible, K:* Kenyapithecus, *A:* Afropithecus, *H:* Heliopithecus, *P:* Prohylabates, *V:* Victoriapithecus.

Sarich and Wilson's dates are broadly correct. Sarich and Wilson's demonstration of an albumin molecular clock in primates and their investigation of its implications has had a profound impact on our understanding of human evolution. This has been discussed by Wilson et al[34] and is illustrated in Figure 5.2.

Most discussion about the implications of Sarich and Wilson's data has focused on the reinterpretation of *Sivapithecus* (including *Ramapithecus*). This and other primate fossils will be discussed in more detail in chapter 8. Briefly, *Sivapithecus* was previously thought to be on the human lineage and to post-date the separation of humans from the other great apes. Its position is now debated, but it is no longer regarded as a direct ancestor of humans. Similarly *Proconsul,* which was thought to be on the human lineage is now generally thought to be a common ancestor of all apes, and *Aegyptopithecus,* which was thought to be on the lineage leading to the nonhuman great apes, is now thought to be ancestral or a sister group to all living catarrhines. It is also now thought to be older than previously estimated. In addition to the reinterpretation of these fossils, a number of new fossil finds have contributed to this revised interpretation. These include *Ouranopithecus,* the Samburu mandible, *Kenyapithecus, Afropithecus, Heliopithecus, Prohylabates* and *Victoriapithecus.* The current interpretation of these fossils in relation to catarrhine phylogeny is shown in Figure 5.2B, and they are discussed in more detail in chapter 8.

The reinterpretation of *Sivapithecus,* in particular, was part of the general acceptance of a much closer relationship between humans and the other African apes. The most widely accepted view currently is that *Sivapithecus* is on the orangutan lineage. However, since *Sivapithecus* first appeared more than 12 Ma ago (see chapter 8), this interpretation is inconsistent with Sarich and Wilson's estimate of an 8 Ma ago divergence of orangutans from African apes. In chapter 10 we will show that it is also inconsistent with rates of DNA evolution.

Sarich and Wilson were unable to resolve the branching order of humans, chimpanzees and gorillas from their immunological analysis of albumins. This was achieved much later through the comparative analysis of DNA. It set off another round of intense controversy.

DNA-DNA HYBRIDIZATION

Several workers have estimated the extent of genomic difference among primate species using DNA-DNA hybridization. We have already commented on Kohne's[35,36] analysis and on the inadequacies of his conclusion of a slowdown of rate in the lineage leading to humans. We pointed out that had Kohne adopted a relative approach to comparing rates he would have arrived at the opposite conclusion. Similarly, if the genomic differences reported by Benveniste[37] are compared in a relative way no difference among primate taxa is evident. His comparisons of humans and Old World monkeys with New World monkeys only are shown in Table 5.2, since these comparisons have not been included in later studies.

The similarity between the human and Old World monkey comparisons is striking. In both cases the mean ΔT_m is 13.1. Benveniste's[37] comparisons also indicate a lack of rate variation between humans and other apes, between colobine and cercopithecoid Old World monkeys, among different species of New World monkeys and humans, Old World monkeys and New World monkeys.

In another study,[1] rate variation was indicated among members of some of the higher order primate taxa (lemurs, lorises, tarsiers and simians), with lemurs having the slowest rate followed by simians, and with tarsiers and lorises having very similar relatively fast rates. Genomic differences were estimated as $T_{50}H$ and the values obtained (Table 5.3) were well above the limit of reliability (see chapter 4). The proportion of nonreassociated tracer DNA is not reported, but would appear, from the melting curves, to be substantial in many cases, particularly those involving tarsier driver DNA. Thus, at best, these results can be considered as a possible indications of rate variation among these primate taxa.

If the indicated variation exists, its pattern is not that predicted by a generation-time effect. The authors did not even consider differences in generation time as an explanation for the variation, preferring selective explanations, or possibly the random fixation of a mutant affecting overall mutation rate. It is worth noting that comparisons of ape (human), Old World monkey (baboon) and New World monkey

Table 5.2. ΔT_m values for comparisons of humans and Old World monkeys (OWM) with New World monkeys (NWM)

	Driver DNA		
Tracer DNA	**Human**	**Baboon**	**Squirrel monkey**
NWM			
Woolly monkey	13.0	13.0	
Capuchin	13.1	13.4	
Howler monkey	13.0	12.9	
Spider monkey	13.3	13.3	
Squirrel monkey	13.2	13.0	
Owl monkey	13.1	13.4	
OWM			
Macacca arctoides			13.0
Macacca nemestrina			13.1
Human			12.7

Modified and reproduced with permission from RE Benveniste, in RJ McIntyre, ed. Molecular evolutionary genetics. New York: Plenum Press, 1985:359-417.

(spider monkey) taxa were included in the analysis, and that there is no indication of any rate variation among these taxa (Table 5.3)

The primate DNA-DNA hybridization study that evoked most controversy was that of Sibley and Ahlquist.[4] The work was very thorough and included more replicate measurements than any previous study and complete reciprocity of driver and tracer status in all comparisons. The species compared were all catarrhines and therefore had sufficiently similar genomes to obtain reliable estimates of divergence using the method. Sibley and Ahlquist had extensive experience with DNA-DNA hybridization analysis, having applied it in investigations of bird phylogeny over many years, and their primate study appeared to be exemplary.

The controversial aspect of the analysis was the result that it appeared to show that chimpanzees are more closely related to humans than to gorillas. This countered the view that had always prevailed—humans are distinct and belong to a separate lineage to the African apes. Sibley and Ahlquist extended their analysis by increasing the number of replicate comparisons performed[7] and confirmed their earlier result.

The response from critics was a vicious attack on their experimental procedures, and a suggestion that they had acted unethically.[38] Sibley and Ahlquist's results implied that humans do not have a distinct evolutionary position relative to other apes. It is not clear whether the intensity of the attack on their work was motivated by an general inability to accept this implication or an inability to accept that the results countered the conclusion reached by their principal critic based on analysis of chromosome morphology.[39]

In any case Sibley et al[15] reanalyzed the relevant components of their data to take account of these criticisms and showed that their earlier conclusions were valid. Their conclusions were also confirmed by an independent analysis by Caccone and Powell[12] and by two-dimensional protein gel electrophoresis and several analyses of sequence

Table 5.3. ***$\Delta T_{50}H$ for comparisons among primate species***

	Driver DNA			
Tracer DNA	**Human**	**Tarsier**	**Galago**	**Lemur**
Human		24.6	29.3	23.4
Baboon	6.9	–	27.8	25.1
Spider monkey	11.2	26.4	30.7	24.5
Tarsier	26.2		30.2	26.9
Galago	26.7	30.1		22.1
Loris	28.1	28.6	11.4	23.1
Lemur	21.9	24.6	22.2	

Reproduced with permission from TI Bonner et al, Nature 1980; 286-420-423.

variation in both nuclear and mitochondrial genes (see references above). There is no longer any sensible reason for doubting that Sibley and Ahlquist were correct in concluding that chimpanzees are more closely related to humans than to gorillas.

The more complete results of Sibley and Ahlquist are presented above the diagonal in Table 5.4, with Caccone and Powell's results shown below the diagonal for comparison. The distances between all pairs of species are approximately the same in both studies. Relative comparison of rates indicates no variation among lineages.[5] This is evident from inspection of Table 5.4. There is approximate uniformity of the values down each column above the diagonal and of the values along each row below the diagonal. The greater degree of relative variation in the data of Caccone and Powell (below the diagonal) is explained by their having made fewer replicate comparisons; Sibley and Ahlquist's estimates are the more precise of the two.

As reported above (Table 5.3), numerous reciprocal comparisons between catarrhines and New World monkeys show a lack of any rate variation between human and Old World monkey lineages. These results combined with those shown in Table 5.4 provide overwhelming evidence for evolutionary rate constancy among catarrhine lineages.

It is also apparent from Table 5.4 that the ratio of human/chimpanzee to human/orangutan genomic difference is approximately 1.6/3.6 = 0.44 and of human/Old World monkey to human/orangutan is approximately 7.3/3.6 = 2.0. Assuming a divergence time of 16 Ma for humans and orangutans, and using these ratios, Sibley and Ahlquist[4] estimated that the divergence between humans and chimpanzees had occurred approximately 7 Ma ago, and that the divergence between apes and Old World monkeys had occurred approximately 32 Ma ago. As we will discuss in chapter 8, there is no fossil evidence of a divergence between humans and orangutans as early as 16 Ma, and we will

Table 5.4. DNA-DNA hybridization comparisons among catarrhines. $\Delta T_{50}H$ from Sibley and Ahlquist (above the diagonal)—values from Caccone and Powell (below the diagonal)

	Human	Chimp	Gorilla	Orangutan	Gibbon	OWM
Human	-	1.6	2.3	3.6	4.8	7.3
Chimp	1.6	-	2.2	3.6	4.8	7.2
Gorilla	2.5	2.6	-	3.6	4.7	7.2
Orangutan	3.5	3.5	3.6	-	4.8	7.4
Gibbon	5.0	4.7	5.2	4.8	-	7.1
OWM	6.8	7.0	7.1	7.3	7.0	-

Reproduced with permission from Easteal, Mol Biol Evol 1991; 8:115-127.

show in chapter 10 that such early divergences are extremely improbable if not impossible.

Sibley and Ahlquist noted that an assumed divergence of 16 Ma for humans and orangutans gives the same estimate of evolutionary rate that they had obtained from their analysis of bird species.[41] They used this as support for their estimated primate divergence times. Their reasoning, of course, assumes that DNA evolves at the same rate in mammals and birds, which may not be the case. Even if it were the case their estimate of the evolutionary rate in birds involves assumptions of a number of divergence times derived from dubious biogeographical interpretation. These include: (1) The initial radiation of the birds (100-130 Ma ago), and the divergences of: (2) New Zealand wrens and Australian passerines (80 Ma ago); (3) Ostriches and rheas (80 Ma ago); (4) Ostriches and emus (80 Ma ago), (5) Kiwis and emus (40-50 Ma ago) and (6) Old World and New World sub-oscines (75 Ma ago).

The inadequacies of the assumptions made in deriving these divergence times have been discussed by Houde.[42] In the first of these events the largest $\Delta T_{50}H$ value obtained for any bird:bird comparison ($\Delta T_{50}H$ = 25-28) is assumed to be equivalent to that between the most divergent bird taxa. However, the taxa actually being compared are the kingfishers, which first appear in the fossil record in the Oligocene, and the passerines (perching birds), which first appear in the Eocene.

Of the remaining five events, ostriches, emus and rheas were presumed to have diverged when Africa, South America and Australia became widely separated. Houde[41] has discussed how the occurrence of ratite (ostriches, emus, etc.) fossils in the Northern Hemisphere[42] undermines the assumption that these species separated at the time of continental separation. Kiwis and emus were presumed to have diverged much later than the separation of Australia from New Zealand. The choice of a particular date after the separation is arbitrary, particularly since it was also assumed that the New Zealand wrens and Australian passerines (which, unlike kiwis, emus and the other ratites, are able to fly), diverged no later than the time of the Australian-New Zealand separation. The Old World and New World sub-oscines, because they can fly, were presumed to have diverged 5 Ma after ostriches and rheas. Although this date may be reasonable, its choice is arbitrary. Any other date after the latter divergence would have been equally reasonable.

In conclusion, Sibley and Ahlquist's demonstration of evolutionary rate uniformity among ape lineages and of the close relationship between chimpanzees and humans are clear and unambiguous. Their interpretation of DNA divergence estimates in terms of species divergence times, however, lacks any solid foundation.

NUCLEAR GENE SEQUENCES

In contrast to the above clear demonstration of evolutionary rate uniformity in catarrhines from DNA-DNA hybridization studies, nucleotide sequence data started to appear during the 1980s that were interpreted as indicating rate variation. In particular, the data were interpreted as demonstrating a slowdown of rate in the lineage leading to humans.[8,25,43,44] Coincidentally, these data have appeared from the same laboratory from which the earlier hominoid slowdown hypothesis was put forward on the basis of comparative analysis of protein sequences.

An important advantage of nucleotide sequences over DNA-DNA hybridization data is that comparisons between species are made in a computer rather than by an experimental laboratory procedure. This obviates the need for reciprocal comparisons and the need for extensive new experimental comparisons when a new species is added to an analysis. Sequence data obtained in different laboratories can be compared without the need for any further experimental work. In fact comparative sequence analysis is now often performed by researchers with little or no involvement in obtaining sequence data. The use of nucleotide sequence data for comparative analysis also overcomes the experimental limitations of the DNA-DNA hybridization method. Reliable estimates of sequence divergence cannot be obtained by DNA-DNA hybridization when genomes differ by more than approximately 15% (see chapter 4). In contrast, reliable estimates of divergence can be obtained between sequences that differ by well over 50%. This means that much more divergent species can be compared using nucleotide sequences than using DNA-DNA hybridization.

Most comparative nucleotide sequence analysis in primates has focused on the ψη-globin gene and the region surrounding it. The results of the most complete comparative analysis of this region,[45] which includes more than 10 kb of sequence in some species, are summarized in Table 5.5.

Some rate variation among taxa is evident from inspection of these data, and variation has been confirmed by formal analysis.[46,47] Particularly notable are the greater rate of substitution in gibbons compared with great apes (relative to Old World monkeys), and in Old World monkeys compared with apes (relative to New World monkeys). Neither of these differences is indicated by the DNA-DNA hybridization studies. Also apparent in the final column of the table is a faster rate of substitution in the galago lineage and a slower rate in the lemur lineage compared with simians. Both of these are consistent with the DNA-DNA hybridization studies, although both are based on relatively short sequences (1.176 and 0.676 kb respectively).

What is *not* evident from the data, despite being claimed by the authors, is a progressive slowdown in the lineage leading to humans. The basis for this claim is the analysis shown in Table 5.6. The time span of various lineages is estimated from assumed divergence times. Thus, for example, strepsirhines are presumed to have diverged from haplorhines 55 Ma ago and galagos are presumed to have diverged from lemurs 44 Ma ago. This gives a time span of 55-44 Ma and

Table 5.5. Percent nucleotide distances between the ψη-globin gene regions of primates (including an outgroup comparison to goat)

	Compared taxon									
Compared taxon	**2.**	**3.**	**4.**	**5.**	**6.**	**7.**	**8.**	**9.**	**10.**	**11.**
1. Human	1.6	1.7	3.4	5.2	7.7	11.4	30.9	36.9	27.6	38.2
2. Chimpanzee		1.8	3.5	5.5	7.8	11.6	31.6	37.1	27.7	38.1
3. Gorilla			3.5	5.3	7.7	11.4	30.9	37.5	28.1	38.3
4. Orangutan				5.4	7.7	11.5	31.5	38.0	29.4	38.8
5. Gibbon					8.6	12.2	32.1	39.5	29.2	40.2
6. Old World Monkey						12.9	32.8	39.1	28.6	41.
7. New World Monkey							32.3	39.2	29.9	40.1
8. Tarsier								38.3	35.1	45.1
9. Galago									32.7	52.4
10. Lemur										33.6
11. Goat										

Modified and reproduced with permission from WJ Bailey et al, Mol Biol Evol 1991; 8:155-184.

hence a length of 11 Ma for the branch leading from the last common ancestor of all living primates to the last common ancestor of lemurs and galagos. The number of substitutions occurring on each branch of the primate phylogeny is derived from a maximum parsimony analysis.[48] The rate of substitution for a particular branch is then obtained by dividing the estimate of the number of substitutions occurring along the branch by the length of the branch.

A clear indication of a progressive slowing of evolutionary rate in the lineage leading to humans emerges from this analysis (Table 5.6). There are, however, two obvious sources of error. First, maximum parsimony is well known as an inappropriate method of estimating branch lengths.[49] Second, the analysis depends on interpretation of the fossil record for estimates of divergence times. Many of the assumed divergence times are almost certainly erroneous. The reader is referred to chapter 8 (particularly Table 8.1) for an account of the primate fossil record.

The adequacy of the analysis can be evaluated in the following way. First, the rate estimates in Table 5.6 can be used to estimate substitution rates in complete lineages leading to humans and different other species. Thus, for example, the rates in Table 5.6 imply that the rate of substitution in the lemur lineage since the last common ancestor of humans and lemurs has been $(4.5 \times 11 + 2.5 \times 44) / (11 + 44) = 2.82$. This is simply the average of the rates in the branches that comprise the lineage weighted by the branch lengths. Similarly, a rate is implied in the human lineage, since the last common ancestor of humans and lemurs, of $(3.5 \times 21 + 1.9 \times 9 + 1.7 \times 6 + 1.1 \times 19) / 55 = 2.21$. The ratio of lemur to human rates is thus $2.82 / 2.21 = 1.28$.

Table 5.6. Variation in substitution rate in the ψη-globin gene region among the branches of primate lineages, as estimated by Bailey et al [46]

Lineage leading to:	Time span (Ma)	Substitution rate
Strepsirhine ancestor	55–44	4.5
Galago	44–0	4.5
Lemur	44–0	2.4
Tarsier	55–0	3.4
Simian ancestor	55–34	3.5
Owl monkey	34–0	2.1
Spider monkey	34–0	1.9
Catarrhine ancestor	34–25	1.9
Rhesus monkey	25–25	1.8
Ape ancestor	25–19	1.7
Gibbon	19–0	1.7
Orangutan	19–0	1.2
Gorilla	19–0	1.2
Chimpanzee	19–0	1.2
Human	19–0	1.1

Reproduced with permission from WJ Bailey et al, Mol Biol Evol 1991; 8:155-184.

Ratios derived in this way are compared with those obtained directly from Table 5.5, in which no assumption is made about species divergence times (Table 5.7).

A number of differences are apparent between the estimates derived from the two different approaches. The most obvious is that the approach of Bailey et al[45] implies a faster evolutionary rate in the lemur lineage than in the human lineage. The data in Table 5.6 imply the opposite. The overall pattern that emerges is that relative to humans, lemurs have the slowest rate followed by the great apes, tarsiers, New World monkeys, gibbons, Old World monkeys and galagos. This is far from a pattern of gradual slowdown associated with proximity to the human lineage.

Bailey et al[45] provided no formal test of rate variation among lineages. However, Li and Tanimura[47] and Li et al[46] had previously applied the formal relative rate test developed by Wu and Li[50] (discussed in more detail in the next chapter) to the 2 kb of ψη-globin gene sequences that were then available for a number of primate taxa, and to sequences from a number of other genes. At the ψη-globin gene, they found significant rate differences between humans and Old World monkeys and between humans and New World monkeys. They found no significant differences between humans and chimpanzees, gorillas or orangutans. Overall they found no significant difference between humans and chimpanzees (5.828 kb), and between humans and orangutans (5.012 kb). They did find significant overall differences be-

Table 5.7. Comparison of estimates of substitution rates in the ψη-globin gene region in different primates relative to the human lineage.

Lineage	Substitution rate relative to human lineage	
	From Table 5.5	**From Table 5.6**
Lemur	0.71	1.28
Galago	2.24	2.04
Tarsier	1.26	1.54
New World monkey	1.27	1.48
Old World monkey	1.48	1.36
Gibbon	1.42	1.54
Orangutan	1.13	1.09
Gorilla	1.11	1.09
Chimpanzee	1.16	1.09

Derived (1) directly from the results presented in Table 5.5; and (2) from the analysis summarized in Table 5.6.

tween humans and gorillas (3.265 kb), humans and Old World monkeys (3.520 kb) and humans and New World monkeys (3.444 kb). They attributed these differences to differences in generation time.

Easteal[51] extended this analysis to include additional sequence data (including the longer sequences from the ψη-globin gene region). He found no significant difference in rate between humans and any other taxon except Old World monkeys (including chimpanzee, gorilla, orangutan, New World monkey and lemur) either for the ψη-globin gene or for the combined sequences. He also pointed out that in both his analysis and that of Li et al[46] the ψη-globin gene region comprised the majority of the sample of nucleotides. Thus the results of overall comparisons, involving a number of genes, would be heavily weighted by the results for this particular gene. A difference in overall rate might thus reflect a difference in this one gene that was not typical of the genome as a whole.

In Li et al's[46] analysis differences in rate were only observed in the ψη-globin gene region. No differences were observed at 11 other loci, although these were all individually much smaller and the human rate was nonsignificantly lower in 28 of 36 separate comparisons with other taxa. In Easteal's expanded analysis, the rate of evolution appeared slower in only 52% of comparisons. Furthermore, when the ψη-globin gene region and its dominant effect were removed from the analysis no significant difference was observed in a combined analysis of the 6 kb that comprised the remaining 18 genes.

Subsequently, Seino et al[52] compared insulin gene sequences and reported a significantly higher rate of substitution in Old World monkeys relative to humans. The difference was great enough that it resulted in an overall difference in rate across genes even when the ψη-globin gene region was omitted from the analysis. However, they published no alignment of their sequences and in repeating their analysis one of us (S.E.) has been unable to find the difference in rate that they reported. Furthermore analysis of an extensive set of additional sequences clearly shows no difference in rate between humans and Old World monkeys (S.E. unpublished results).

Two other recent studies have given further support to rate constancy among simian lineages. Hayasaka et al[53] compared the rates of evolution in the intergenic region between the γ-globin and ε-globin genes of various primates and rabbits. The results are shown in Table 5.8. Although no formal relative rate test was performed, it is apparent from the table that there is no significant difference in evolutionary rate of this approximately 5 kb sequence between humans and New World monkeys relative to tarsiers, galagos or rabbits. There is an indication of a faster rate of evolution in tarsiers and galagos relative to both humans and New World monkeys. The substitution rate in the tarsier and galago lineages relative to the human rates are respectively 1.28 and 2.0, which are similar to the values obtained from analysis of the ψη-globin gene (Table 5.7), suggesting that these differences may be real.

Kawamura et al[54] compared 1.36 kb of noncoding sequence of the immunoglobulin α gene in humans, gorillas, chimpanzees, orangutan and crab-eating macaque. Their results are shown in Table 5.9. They found no variation in rate among the ape species, using the macaque sequence for reference. They were not able to do a formal relative rate test between the apes (including humans) and the Old World monkey species (macaque) as they did not have a suitable outgroup reference sequence. They used a less direct method of comparing the evolutionary rates in apes and Old World monkeys. They constructed a phylogenetic tree and from this they estimated the number of nucleotide substitutions per site in the branches leading to orangutans and macaques as 0.0270 and 0.0302 respectively. The ratio of these two values is 1.12.

Table 5.8. Percent nucleotide sequence divergence between humans, New World monkeys, tarsiers and galagos relative to tarsiers and galagos and rabbits in the intergenic region between γ- and ε-globin genes

	Reference species		
	Tarsier	**Galago**	**Rabbit**
Human	32	39	42
Capuchin (NWM)	31	39	43
Tarsier		43	49
Galago			55

Reproduced with permission from K Hayasaka et al, Genomics 1993; 18:20-28.

They compared this ratio with the comparable ratio obtained for the two γ-globin genes (1.6 and 1.74) the δ-globin gene (1.69) and the pseudo ψη-globin gene (2.37). This showed that relative to Old World monkeys, the substitution rate in apes is lower for this gene than for these others. It can be seen from Table 5.4 that the comparable ratio obtained from DNA-DNA hybridization data is approximately 2.0. Since the relative rate tests described above show that there is no difference in rate between apes and Old World monkeys for the δ- and γ-globin genes[54] or for the DNA-DNA hybridization data,[55] this implies that, if anything, the substitution rate at the immunoglobulin α gene is faster in the ape (including human) lineage than in the Old World monkey lineage.

Clearly, if there are any differences in the evolutionary rate of DNA among catarrhine lineages they are not general in nature since no differences in rate are evident in the DNA-DNA hybridization data. There is no evidence of any rate variation among ape lineages. There does appear to be some difference in rate in the region of the ψη-globin gene between apes and Old World monkeys. There is no evidence that this difference extends beyond this region, and there is substantial evidence that it does not. There may also be an even smaller difference in rate between apes and New World monkeys. Again, the available data indicate that this difference does not extend beyond the ψη-globin gene region. There are indications, both from DNA-DNA hybridization data and comparative sequence analysis of variable rates among strepsirhines. Lemurs appear to evolve more slowly than simians and galagos and lorises appear to evolve more rapidly. The distances between these taxa are, however, too great to put much reliance on the DNA-DNA hybridization results, and the sequence results are all from one region of the genome (the β-globin gene family). Additional sequence data are needed to confirm whether real differences exist. The history of analysis of the ψη-globin gene illustrates the fallacy of basing general conclusions on results from only one gene or genomic region. One thing is clear, however; whatever overall differences might exist, they cannot be explained by differences in generation time.

Table 5.9. Nucleotide distances between the immunoglobulin α genes of catarrhine species

	Human	Chimpanzee	Gorilla	Orangutan
Chimpanzee	2.8 ± 0.5			
Gorilla	2.5 ± 0.4	1.9 ± 0.4		
Orangutan	5.6 ± 0.7	5.2 ± 0.6	4.6 ± 0.6	
Macaque	6.4 ± 0.7	6.0 ± 0.7	5.9 ± 0.7	5.8 ± 0.7

Reproduced with permission from S Kawamura et al, Mol Biol Evol 1991; 8:743-752.

MITOCHONDRIAL DNA

The results of comparative analysis of mitochondrial genome sequences confirm a lack of molecular evolutionary rate variation among simian taxa.[51] A 0.9 kb region of the mitochondrial genome has been sequenced in 12 primate species.[56,57] There are no significant differences in the rates of transversion-type substitutions among species (Table 5.10). Transition-type substitutions are omitted because of the possibility that real differences between lineages might be masked by the effects of substitution saturation.

There is also no significant difference in rate in the evolution of the mitochondrial cytochrome c oxidase subunit II genes of humans and Old World monkeys using the cow gene as an outgroup reference.[52]

More recently Horai et al[58] have reported the sequences of a 4.8 kb region of the mitochondrial genome of common and pygmy chimpanzees, gorillas, orangutans, and siamangs and have compared these to the human sequence. The most significant result of this comparison is the overwhelming support for the closer affinity of chimpanzees and humans than of either species with gorillas. Tajima[59] applied a new method of testing for rate constancy to these data. His analysis, which included all sites (synonymous and nonsynonymous, as well as those at tRNA genes) indicated a significant slower rate in pygmy chimpanzees than in the other great apes, relative to siamangs. Otherwise, however, no differences were observed. The results (omitting pygmy chimpanzees) for synonymous differences in protein-coding genes and in tRNA genes are shown in Table 5.11. In none of the comparisons is the human rate the lowest.

In conclusion, Sarich and Wilson's proposal of a molecular clock for primates is overwhelmingly supported in simians by subsequent

Table 5.10. Transversion differences among primates in a 0.9 kb region of the mitochondrial genome

	Compared taxa						
Compared taxa	**Chimp**	**Gorilla**	**Orangutan**	**Gibbon**	**OWM**	**NWM**	**Strepsirhine**
Human	0.6	0.9	3.9	5.0	8.3	11.0	15.5
Chimpanzee		1.0	3.8	4.9	8.2	11.1	15.5
Gorilla			3.7	5.0	8.1	10.8	15.1
Orangutan				5.8	8.5	12.4	14.5
Gibbon					8.6	11.7	15.0
Old World monkey						11.4	14.6
New World monkey							15.9

Reproduced with permission from S Easteal, Molecular Biology and Evolution 1991; 8:115-127.

Table 5.11. Sequence differences of a 4.8 kb region of the mitochondrial genome in ape species

	Compared species				
Compared species	**Human**	**Chimp**	**Gorilla**	**Orangutan**	**Siamang**
Human		326	409	477	543
Chimpanzee	28		376	474	530
Gorilla	41	34		489	544
Orangutan	65	61	64		528
Siamang	72	62	69	78	

Synonymous differences in protein coding regions are shown above the diagonal; differences in tRNA genes are shown below the diagonal.
Modified and reproduced with permission from S Horai, J Mol Evol 1992; 35:32-43.

DNA analysis. It remains to be seen if it also applies to strepsirhines. Although the general rate of DNA evolution does not vary among simian lineages, the rates of some specific genes or gene regions clearly do vary. These, however, are exceptions. What causes these exceptions is not clear. They may result from differences in local mutation rates or from the effects of natural selection. In either case they illustrate the danger of drawing general conclusions about evolutionary rates from the analysis of individual genes. Because any rate differences that do exist are local rather than general they cannot be explained by differences in the general characteristics of the different lineages such as generation time, metabolic rate or any other factor that might affect global mutation rates. The general pattern of rate uniformity also means that the relative divergence times of primate taxa can be estimated from molecular data. We shall return to this last point in chapter 10.

References

1. Bonner TI, Heinemann R, Todaro GJ. Evolution of DNA sequences has been retarded in Malagasy primates. Nature 1980; 286:420-423.
2. Goodman M, Tagle DA, Fitch DHA, Bailey W, Czelusniak J, et al. Primate evolution at the DNA level and a classification of hominoids. J Mol Evol 1990; 30:260-266.
3. Miyamoto MM, Goodman M. DNA systematics and the evolution of primates. Ann Rev Ecol Syst 1990; 21:197-220.
4. Sibley CG, Ahlquist JE. The phylogeny of the hominoid primates, as indicated by DNA-DNA hybridization. J Mol Evol 1984; 20:2-15.
5. Felsenstein J. Estimation of hominoid phylogeny from a DNA hybridization data set. J Mol Evol 1987; 26:123-131.
6. Goldman D, Giri PR, O'Brien S. A molecular phylogeny of the hominoid primates as indicated by two-dimensional protein electrophoresis. Proc Natl Acad Sci (USA) 1987; 84:3307-3311.
7. Sibley CG, Ahlquist JE. DNA hybridization evidence of hominoid phy-

logeny: Results of an expanded data set. J Mol Evol 1987; 26:99-121.
8. Fitch DHA, Mainone C, Slightom JL, Goodman M. The spider monkey ψη-globin gene and surrounding sequences: recent or ancient insertions of LINEs and SINEs? Genomics 1988; 3:237-255.
9. Holmquist R, Miyamoto MM, Goodman M. Analysis of higher-primate phylogeny from transversion differences in nuclear and mitochondrial DNA by Lake's methods of evolutionary parsimony and operator metrics. Mol Biol Evol 1988; 5:217-236.
10. Miyamoto MM, Koop BF, Slightom JL, Goodman M, Tennant MR. Molecular systematics of higher primates: geneological relations and classification. Proc Natl Acad Sci USA 1988; 85:7627-7631.
11. Koop BF, Tagle DA, Goodman M, Slightom JL. A molecular view of primate phylogeny and important systematic and evolutionary questions. Mol Biol Evol 1989; 6:580-612.
12. Caccone A, Powell JR. DNA divergence among hominoids. Evolution 1989; 43:925-942.
13. Goodman M, Koop BF, Czelusniak J, Fitch DH, Tagle DA, et al. Molecular phylogeny of the family of apes and humans. Genome 1989; 31:316-335.
14. Williams SA, Goodman M. A statistical test that supports a human / chimpanzee clade based on noncoding DNA sequence data. Mol Biol Evol 1989; 6:325-330.
15. Sibley CG, Comstock JA, Ahlquist JE. DNA hybridization evidence of hominoid phylogeny: a reanalysis of the data. J Mol Evol 1990; 30:202-236.
16. Ruvolo M, Disotell TR, Allard MW, Brown WM, Honeycutt RL. Resolution of the African hominoid trichotomy by use of a mitochondrial gene sequence. Proc Natl Acad Sci (USA) 1991; 88:1570-1574.
17. Horai S, Satta Y, Hayasaka K, Kondo R, Inoue T, et al. Man's place in hominoidea revealed by mitochondrial DNA genealogy. J Mol Evol 1992; 35:32-43.
18. Margoliash E. Primary structure and evolution of cytochrome *c*. Proc Natl Acad Sci USA 1963; 50:672-679.
19. Sarich VM, Wilson AC. Rates of albumin evolution in primates. Proc Natl Acad Sci (USA) 1967; 58:142-148.
20. Goodman M. Evolution of the immunologic species specificity of human serum proteins. Hum Biol 1962; 34:104-150.
21. Goodman M. Serological analysis of the systematics of recent hominoids. Hum Biol 1963; 35:377-436.
22. Goodman M, Moore GW, Matsuda G. Darwinian evolution in the genealogy of haemoglobin. Nature 1975; 253:603-608.
23. Czelusniak J, Goodman M, Hewett-Emmett D, Weiss ML, Venta PJ, et al. Phylogenetic origins and adaptive evolution of avian and mammalian haemoglobin genes. Nature 1982; 398:297-300.
24. Goodman M, Braunitzer G, Stangl A, Schrank B. Evidence on human origins from haemoglobins of African apes. Nature 1983; 303:546-548.

25. Goodman M. Rates of molecular evolution: the hominoid slowdown. BioEssays 1985; 3:9-14.
26. Tateno Y, Nei M. Goodman et al.'s method for augmenting the number of nucleotide substitutions. J Mol Evol 1978; 11:67-73.
27. Holmquist R. The method of parsimony: an experimental test and theoretical analysis of the adequacy of molecular restoration studies. J Mol Evol 1979; 135:939-958.
28. Kimura M. Was globin evolution very rapid in its early stages?: a dubious case against the rate-constancy hypothesis. J Mol Evol 1981; 17:110-113.
29. Kimura M. The neutral theory of molecular evolution. Cambridge: Cambridge University Press, 1983.
30. Sarich VM, Wilson AC. Quantitative immunochemistry and the evolution of primate albumins: micro-complement fixation. Science 1966; 154:1563-1566.
31. Wasserman E, Levine L. Quantitative micro-complement fixation and its use in the study of antigenic structure by specific antigen-antibody inhibition. J Immunol 1961; 87:290-295.
32. Sarich VM, Wilson AC. Immunological time scale for hominid evolution. Science 1967; 158:1200-1203.
33. Maxson RD, Maxson LR. Micro-complement fixation: a quantitative estimator of protein evolution. Mol Biol Evol 1986; 3:375-388.
34. Wilson AC, Ochman H, Prager EM. Molecular time scale for evolution. Trends Genet 1987; 3:241-247.
35. Kohne DE. Evolution of higher-organism DNA. Quart Rev Biophys 1970; 3:327-375.
36. Kohne DE, Chiscon JA, Hoyer BH. Evolution of primate DNA sequences. J Hum Evol 1972; 1:627-644.
37. Benveniste RE. The contributions of retroviruses to the study of mammalian evolution. In: McIntyre RJ, ed. Molecular evolutionary genetics. New York: Plenum Press, 1985:359-417.
38. Marks J, Schmid CW, Sarich VM. DNA hybridization as a guide to phylogeny: relations of the hominoidea. J Hum Evol 1988; 17:769-786.
39. Marks J. Hominoid cytogenetics and evolution. Phys Anthrop Yrbk 1983; 25:125-153.
40. Sibley CG, Ahlquist JE. The phylogeny and relationships of the ratite birds as indicated by DNA-DNA hybridization. In: Scudder GGE, Reveal JL, eds. Evolution today. Proc Second Int Cong Syst Evol Biol, 1981:301-335.
41. Houde P. Critical evaluation of DNA hybridization studies in avian systematics. Auk 1987; 104:17-32.
42. Houde P. Ostrich ancestors found in the Northern Hamisphere suggest new hypothesis of ratite origins. Nature 1986; 324:563-565.
43. Goodman M, Koop BF, Czelusniak J, Weiss ML. The η-globin gene; its long evolutionary history in the β-globin gene family of mammals. J Mol Biol 1984; 180:803-823.
44. Koop BF, Goodman M, Xu P, Chan K, Slightom JL. Primate η-globin

DNA sequences and man's place among the great apes. Nature 1986; 319:234-237.

45. Bailey WJ, Fitch DHA, Tagle DA, Czelusniak J, Slightom JL, et al. Molecular evolution of the ψη-globin gene locus: gibbon phylogeny and the hominoid slowdown. Mol Biol Evol 1991; 8:155-184.
46. Li W-H, Tanimura M, Sharp PM. An evaluation of the molecular clock hypothesis using mammalian DNA sequences. J Mol Evol 1987; 25:330-342.
47. Li W-H, Tanimura M. The molecular clock runs more slowly in man than in apes and monkeys. Nature 1987; 326:93-96.
48. Moore GW, Barnabas J, Goodman M. A method for constructing maximum parsimony ancestral amino acid sequences on a given network. J Theor Biol 1973; 38:459-485.
49. Nei M. Molecular evolutionary genetics. New York: Columbia University Press, 1987.
50. Wu C-I, Li W-H. Evidence for higher rates of nucleotide substitution in rodents than in man. Proc Natl Acad Sci (USA) 1985; 82:1741-1745.
51. Easteal S. The relative rate of DNA evolution in primates. Mol Biol Evol 1991; 8:115-127.
52. Seino S, Bell GI, Li W-H. Sequences of primate insulin genes support the hypothesis of a slower rate of molecular evolution in humans and apes than in monkeys. Mol Biol Evol 1992; 9:193-203.
53. Hayasaka K, Skinner CG, Goodman M, Slightom JL. The γ-globin genes and their flanking sequences in primates: findings with nucleotide sequences of capuchin monkey and tarsier. Genomics 1993; 18:20-28.
54. Kawamura S, Tanabe H, Watanabe Y, Kurosaki K, Saitou N, et al. Evolutionary rate of immunoglobulin alpha noncoding region is greater in hominoids than in Old World monkeys. Mol Biol Evol 1991; 8:743-752.
55. Catzeflis F, Sheldon FH, Ahlquist JE, Sibley CG. DNA-DNA hybridization evidence of the rapid rate of muroid rodent DNA evolution. Mol Biol Evol 1987; 4:242-253.
56. Brown WM, Prager EM, Wang A, Wilson AC. Mitochondrial DNA sequences of primates: tempo and mode of evolution. J Mol Evol 1982; 18:225-239.
57. Hayasaka K, Gojobori T, Horai S. Molecular phylogeny and evolution of primate mitochondrial DNA. Mol Biol Evol 1988; 5:626-644.
58. Horai S, Satta Y, Hayasaka K, Kondo R, Inoue T, et al. Man's place in Hominoidea revealed by mitochondrial DNA genealogy. J Mol Evol 1992; 35:32-43.
59. Tajima F. Simple methods for testing the molecular evolutionary clock hypothesis. Genetics 1993; 135:599-607.

CHAPTER 6

The Order of Orders: Molecular Evolution in Placental Mammals

In comparisons of molecular evolutionary rates between humans and other mammals it has almost always been the human lineage in which the rate is claimed to have been slower. In the previous chapter we discussed this issue in relation to comparisons with other primates, and suggested that proposed rate differences were more a reflection of inadequate analysis than of anything real. In this chapter we discuss comparisons between primates (particularly humans) and other mammals. As was the case for the analysis within primates, both the relative approach and approaches that depend on paleontological interpretation have been applied; and, as in primates, the two different approaches have provided different answers. The relative approach could be readily applied within primates because of a well established branching order. This is not the case for comparisons among the orders of placental mammals, and the order in which placental orders diverged has been an important issue in the debate about evolutionary rates.[1-3]

There has been particular interest in comparing evolutionary rates in rodents and primates both because of the relative abundance of available sequence data and because of the very substantial difference in generation time that exists between species in these two orders. This makes the comparison very useful in investigating the role of generation time (and more recently metabolic rate) in determining the rate of molecular evolution.

TESTS INVOLVING PALEONTOLOGICAL INTERPRETATION

One of the best known and most widely cited studies purporting to show variation in rate among lineages (including a faster rate of evolution in rodents than in primates) is that of Britten.[4] He compiled genetic distances from DNA-DNA hybridization studies and nucleotide sequence comparisons across a range of animal taxa. Then, as summarized in Figure 6.1, he plotted genetic distance as a function of divergence time and concluded, among other things, that rodents had evolved approximately five times faster than higher primates. Our concern here is with mammals, so that his comparisons involving echinoderms and *Drosophila* are omitted from Figure 6.1. It can be seen from the figure that Britten's conclusion depends very much on a single assumed divergence time—that of mice and rats 10-25 Ma ago. We will show in chapters 7 and 10 that this assumption is almost certainly erroneous. In fact we will show that most of the divergence time estimates made by Britten are improbable, but the mouse-rat divergence is the most important in the context of his conclusion of a 5-fold greater rate of DNA evolution in the rodent lineage.

Other studies have reached the same conclusion based on much the same assumptions about divergence times. Brownell[5] and Catzeflis et al[6] both concluded, on the basis of DNA-DNA hybridization data and an assumed divergence of mice from rats of approximately 10 Ma, that the DNA of rodents has evolved ten times faster than that of primates. Li and Tanimura[7] and Li et al[8] assumed a 15 Ma divergence for mice and rats, a 7 Ma divergence for humans and chimpanzees and a 25 Ma divergence for humans and Old World monkeys. They concluded that the rodent rate is 4-10 times greater than the higher primate rate. In a similar vein, Holmes,[9] by assuming that primates and artiodactyls separated 70 Ma ago, estimated that a molecular clock implied a rodent separation 108.9 Ma ago and a mouse-rat separation 33.8 Ma ago. Since she considered these dates too early, she was led to conclude that rodent genes show elevated rates of evolution. Bulmer et al[10] assumed a 12 Ma divergence for mice and rats and concluded that there was a 5-fold greater increase in rate in the rodent lineage since this divergence. Interestingly they suggest that before the divergence, the rate of evolution in rodents was much the same as it had been in artiodactyls and primates. This implies a slower rate in critedids (hamsters etc.) than in murids, which is refuted by the later work of O'hUigin & Li[11]).

What all these studies actually demonstrate is that *either* evolutionary rates vary *or* the fossil record has been misinterpreted; a relative rate test is required to distinguish between these two possibilities. We discuss the fossil record in more detail in coming chapters. It is worth noting here, however, that the molecular systematics of murid rodents suggests a substantial misinterpretation of their meager fossil record. This issue is discussed by Wilson et al[12] and is illustrated in Figure 6.2. Briefly, the fossil record of murid and related rodents consists mainly of molar teeth, and taxonomy is based largely on the cusp pattern of these. This includes the fossils thought to give evidence of the divergence time of mice and rats. The molar teeth evidence indicates a closer relationship between true mice and spiny mice than between true mice and rats. The molecular evidence, on the other hand, shows clearly that spiny mice are distantly related to both true mice and rats; they are in fact more closely related to gerbils.[13] This implies that the cusp patterns of molar teeth have evolved convergently and

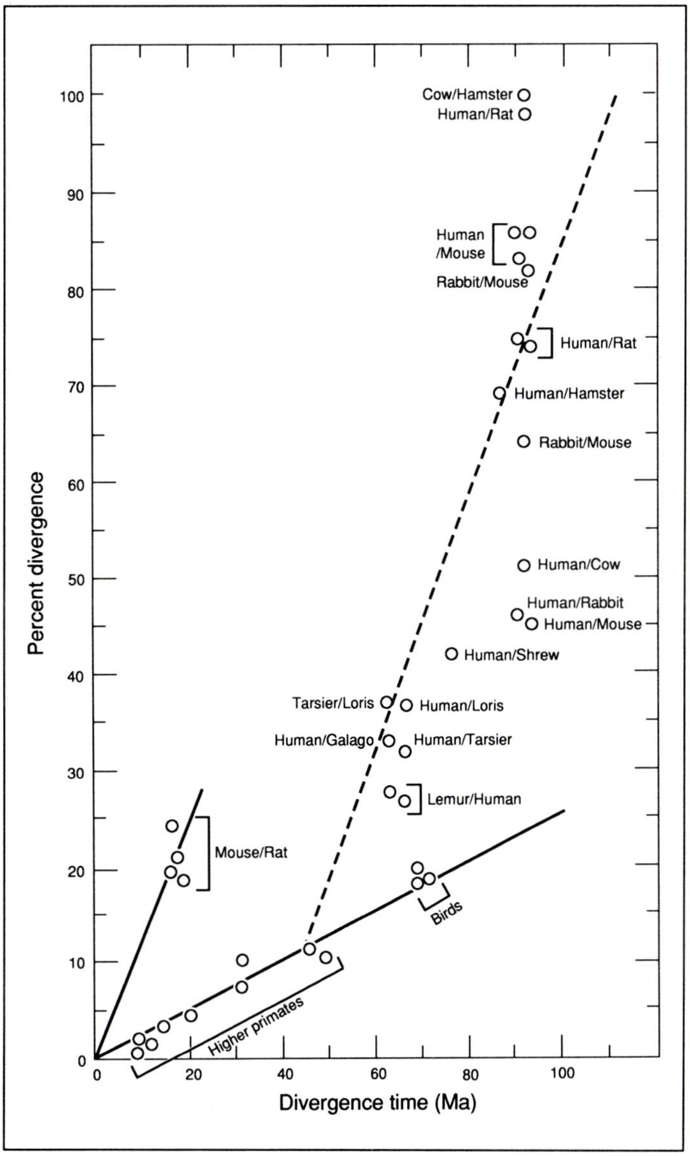

Fig. 6.1. Percent sequence divergence of vertebrate taxa as a function of their divergence times as estimated by Britten. Modified and reprinted with permission from RJ Britten, Science 1986; 231:1393-1398.

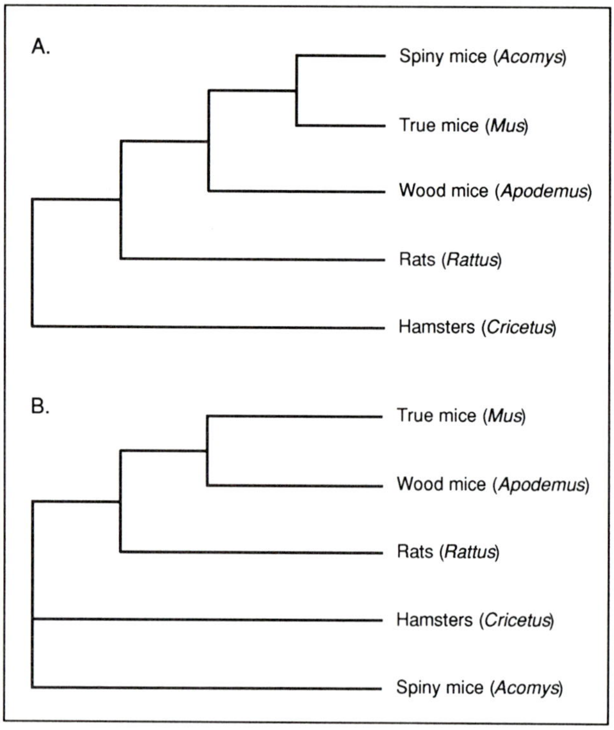

Fig. 6.2. The relationships of the rodents based on: (A) evidence from paleontology and anatomy; and (B) immunological comparisons of albumin, transferrin and lysozyme. Modified and reprinted with permission AC Wilson et al, Trends Genet 1987; 3:241-247.

are unreliable taxonomic characters. They cannot, therefore, provide reliable evidence of species divergence times, including the divergence times of mice and rats.

EARLY STATISTICAL TESTS

The first attempt at a formal statistical approach to testing of molecular evolutionary rate variation was introduced by Ohta and Kimura.[14] They compared the expected and observed variance in the rate of amino acid substitution for α- and β-globins and for cytochrome c among a number of different vertebrate taxa. They concluded that "the variations in evolutionary rates among highly evolved animals are larger than expected from chance". This was contrary to the expectation of their newly developed neutral theory. They explained this away by claiming that the differences in rate were small and by suggesting that they might be the result of the action of slightly deleterious selection. In fact the apparent differences in rate are probably better explained as resulting from errors in their assumptions about the divergence times of the sequences they were comparing. For example they assumed that all mammalian orders they compared diverged approximately 80 Ma ago, which, as we show, now seems unlikely. They also assumed that teleost and elasmobranch fish diverged from mammals at the same time, which is not supported by the fossil record.[15,16] In addition, their comparisons of artiodactyl β-globin genes probably involved paralogous rather than orthologous comparisons, as the evolutionary relationships of this gene family in artiodactyls is complex.[17]

Langley and Fitch[18] and Fitch and Langley[19] concluded that rates of amino acid substitution varied among mammalian lineages, based on their likelihood ratio test for rate constancy. There are, however, two main problems with their analysis which make their results questionable. The first is that their estimates of nucleotide substitution rates may be inaccurate since they were derived from amino acid rather than nucleotide sequences. The second problem is that their assumed sequence phylogeny is probably incorrect in two respects. They assumed that rabbits and rodents are sister taxa which now seems unlikely, and they assumed that rodents diverged from primates later than artiodactyls, which also now seems unlikely. We discuss these phylogenetic issues in more detail later in this chapter. Their method has not been reapplied with these deficiencies corrected, so we cannot be sure that they entirely account for the apparent rate variation.

Kimura[20] developed another method of testing for evolutionary rate uniformity. This involved the estimation of the number of substitutions, d, in several lineages that diverged from each other at approximately the same time (Fig. 6.3), i.e. that have a "star phylogeny". The test involves an estimation of the ratio (R) of the observed variance (V_d) to the expected variance (σ^2_d) in amino acid substitution rate among the lineages. The star phylogeny is important in estimating σ^2_d because it allows the assumption to be made that the d values follow a Poisson distribution for which the variance equals the mean. If the evolutionary rate is uniform among lineages then $R = 1$, and a X^2 distribution can be used to test for departure of R from unity.

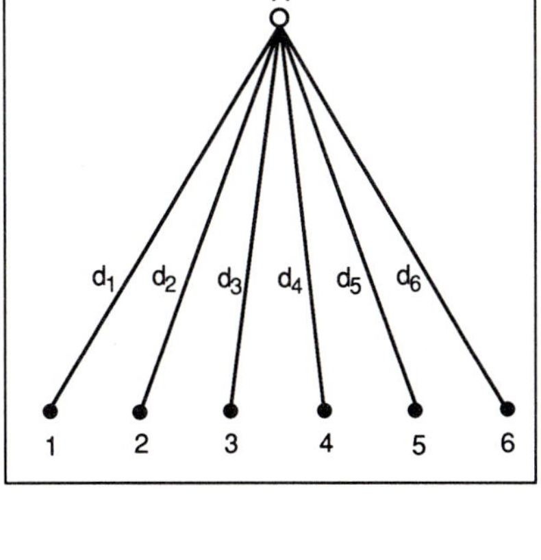

Fig. 6.3. A star phylogeny as assumed in Kimura's[20] test for evolutionary rate uniformity. Estimates are obtained of the number of amino acid substitutions (d_i) occurring in a number of lineages that diverged from each other at approximately the same time. Modified and reproduced with permission from M Kimura. The neutral theory of molecular evolution. Cambridge: Cambridge University Press, 1983.

A. Hemoglobin B

	Human	Mouse	Rabbit	Dog	Horse	Cow
Human		27	14	15	25	25
Mouse			28	30	36	39
Rabbit				21	25	30
Dog					30	28
Horse						30
Cow						

R = 3.1

B. Myoglobin

	Human	Seal	Badger	Horse	Cow	Whale
Human		24	17	19	29	25
Seal			19	20	28	26
Badger				16	25	26
Horse					18	19
Cow						30
Whale						

R = 1.7

C. Cytochrome *c*

	Human	Chicken	Turtle	Snake
Human		13	15	14
Chicken			8	19
Turtle				22
Snake				

R = 3.3

D. Ribonuclease

	Cow	Whale	Horse	Rat
Cow		$\frac{30}{124}$	$\frac{32}{123}$	$\frac{41}{124}$
Whale			$\frac{28}{123}$	$\frac{39}{124}$
Horse				$\frac{39}{123}$
Rat				

R = 2.4

Fig. 6.4. Numbers of amino acid differences between hemoglobin B, myoglobin, cytochrome c and ribonuclease proteins in different lineages. R is the ratio of observed to expected variance in substitution rate among lineages. For a star phylogeny, R = 1 when evolutionary rates are uniform. Modified and reprinted with permission from M Kimura. The neutral theory of molecular evolution. Cambridge: Cambridge University Press, 1983.

The results of Kimura's[20] application of this test to sequences of four proteins are shown in Figure 6.4. In all cases $R > 1$. In the cases of hemoglobin B and cytochrome c, the departure from unity is significant. On the face of it these results would appear to have demonstrated that evolutionary rates vary among lineages, yet Kimura maintained that "although a strict constancy may not hold, yet a rough constancy of the evolutionary rate for each molecule among various lineages is a rule rather than an exception".

Takahata[21] assessed possible neutral departures from the Poisson model of molecular evolution might. He investigated the effects of multiple simultaneous substitutions in the same gene, population bottlenecks and slightly deleterious substitutions, and changes in the degree of selective constraint. He concluded that the elevated variances in substitution rates could be explained under neutral models. Gillespie,[22-24] on the other hand, has proposed complex models of selection to explain the elevated variances. These are described in detail in his book[25] and will not be discussed further here.

All these observed and expected substitution rate variances are discrepant, however, only if the sequences being compared are related by a star phylogeny. If they are not related in this way then Kimura's test and the results derived from it are invalid. Before any conclusions can be drawn from the results of Kimura's test, the assumption of a star phylogeny must be tested. We show below that for the orders of placental mammals, the focus of most of Kimura's tests, the assumption is not valid.

RELATIVE RATE TESTS

The principle of the relative approach to testing for evolutionary rate uniformity among lineages was described in the previous chapter. An important advance in its application was the development by Wu and Li[26] of a formal statistical test on the assumption that the Poisson process gives a reasonable model of neutral evolution.

Wu and Li[26] compared evolutionary rates in rodents and humans using a variety of species, including dogs, rabbits, pigs, cows and goats, as reference species. They found significant rate variation in 8 of 24 comparisons of synonymous substitutions, and in 4 of 24 nonsynonymous comparisons. Overall they found that the rates of synonymous and nonsynonymous substitution in the rodent lineage were two times and 1.3 times greater respectively than in the primate lineage. They also found that the rates in 5' and 3' untranslated regions of genes were more than twice as great in the rodent lineage compared with the primate lineage. It should be remembered that comparisons of evolutionary rates between rodents and primates based on fossil record estimates of divergence times indicated a 5- to 10-fold greater rate in rodents.

Easteal[1] pointed out that Wu and Li's result depended entirely on their assumed phylogeny, specifically on their assumption that rodents

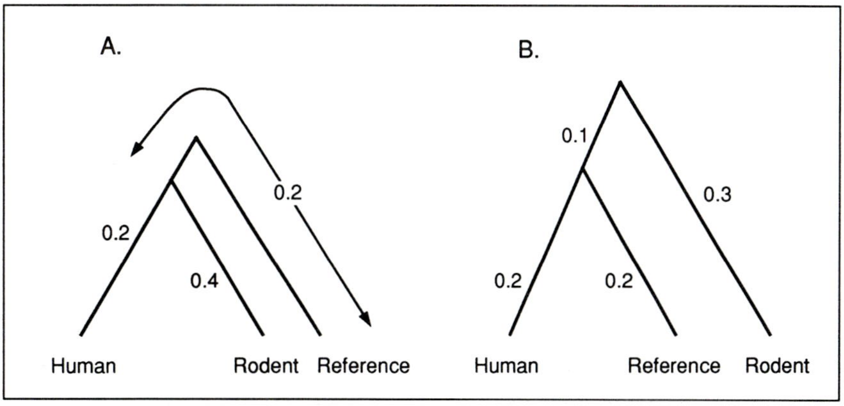

Fig. 6.5. A: Phylogeny assumed by Wu and Li[26] in their test of evolutionary rate uniformity between human and rodent lineages showing the implied number of substitutions per 4-fold degenerate site. Under this phylogeny there is a 2-fold difference in rate between humans and rodents. B: An alternative phylogeny that explains the differences in substitution rate by differences in branch length. Modified and reprinted with permission from Easteal, Mol Biol Evol 1985; 2:450-453.

are more closely related to humans than are the species used for reference. He showed that under the reverse phylogeny (i.e. one in which the reference species are more closely related to humans than are rodents) variation in substitution rate among the three branches can be explained by differences in branch length (Fig. 6.5). In other words, if the reverse phylogeny were the correct one then there might be no rate variation among the lineages.

Easteal[27] investigated this further by estimating the phylogeny of rodents, artiodactyls (cows and goats), lagomorphs (rabbits) and primates from their globin gene sequences. He found that among the members of the β-globin gene family, the adult-expressed and embryonic-expressed genes both indicated with high probability a primary branch of rodents followed by artiodactyls, then lagomorphs and primates. The phylogenies were rooted by paralogous genes, i.e. the adult-expressed phylogeny was rooted by the embryonic expressed genes and *vice versa*. This phylogeny was the opposite of that assumed by Wu and Li,[26] suggesting that their finding of a faster rate of substitution in rodents relative to primates was invalid.

Easteal[27] also showed that there is no evidence of rate variation among the four lineages in either adult or embryonic expressed genes, using paralogues for reference. Similarly he showed a lack of evidence for rate variation between primates and lagomorphs using artiodactyls for reference, and between primates and artiodactyls, and artiodactyls and lagomorphs using rodents for comparison.

The relatively early divergence of rodents relative to the other placental mammals has been confirmed by using marsupial sequences as outgroups[28] (Fig. 6.6), by analysis of the sequences of a larger number genes,[29] and by comparison of mitochondrial genomes,[30] although the relative branching order of primates, artiodactyls and lagomorphs is not as easily resolved as first appeared[10] (Fig. 6.7). An important aspect of these phylogenies is that they clearly show that the different mammalian orders did not diverge from each other at approximately the same time. In other words, the star phylogeny model is incorrect; this is discussed in more detail by Graur.[31] It means that the Kimura's[20] test of rate constancy cannot be applied to sequences from different mammalian orders and that the results of its application in this context should be disregarded.

The early divergence of rodents means that species belonging to most (if not all) other placental orders cannot be used as a reference in relative rate tests between rodents and primates. Rodents can, however, be used as a reference to compare many other placental orders. When this has been done[10,27,28] there is no evidence for rate differences among primates, lagomorphs and artiodactyls, and this has been confirmed by using marsupial[28] and bird[29] sequences for reference. The primate-lagomorph comparison is of particular interest in the context of discussion about the role of generation time in determining evolu-

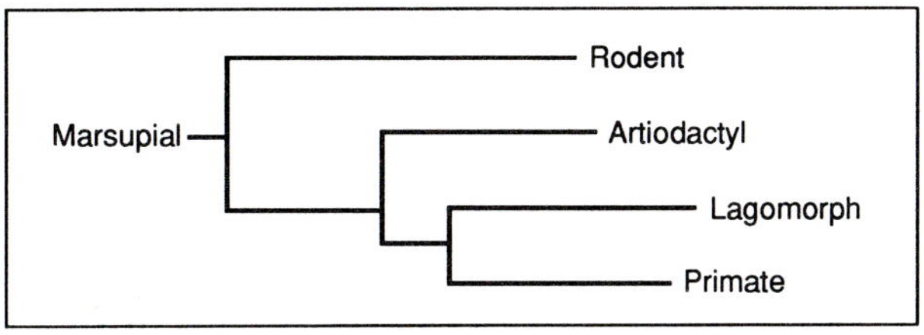

Fig. 6.6. Relationships of four eutherian orders, primates, lagomorphs, artiodactyls and rodents, relative to marsupials. The maximum-likelihood tree is based on coding and noncoding sequences of the globin genes. Modified and reprinted with permission from S Easteal, Genetics 1990; 124:165-173.

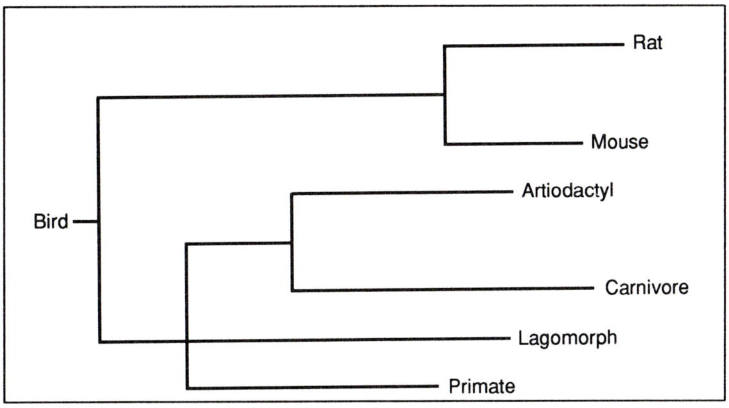

Fig. 6.7. Phylogenetic relationships of mammalian taxa based synonymous substitutions of 58 genes and rooted to chicken. Branches are drawn to scale and represent the proportion of synonymous substitutions per site along each branch. Modified and reprinted with permission from M Bulmer et al, Proc Natl Acad Sci (USA) 1991; 88:5974-5978.

tionary rates. Lagomorphs and primates have very different generation times and the finding that they do not have different rates of molecular evolution should be sufficient to refute suggestions that generation time is important in determining molecular evolutionary rates. This, however, has not been the case and the issue has continued to be disputed.

Easteal[28] compared substitution rates in four genes among primates, rodents, lagomorphs and artiodactyls using marsupials as an outgroup. Overall he found no differences in rate at synonymous sites and in noncoding regions. For nonsynonymous sites, however, he found a 1.65 times greater rate in rodents than in artiodactyls and a 1.76 times greater rate in rodents compared with primates. Using bird sequences for reference Li et al[29] found a 1.5 times greater nonsynonymous substitution rate in rodents compared with primates and Gu and Li,[32] analyzing a larger sample of genes, found a 1.4 times greater rate in rodents. Easteal and Collet,[33] using marsupial sequences for reference also found a 1.4 times greater rate of nonsynonymous substitution in rodents compared to primates. At 4-fold degenerate sites and in noncoding DNA, however, they found no significant difference between the lineages.

The results of comparisons of molecular evolutionary rate between primates and rodents are summarized in Table 6.1. In studies involving fossil-derived divergence times a 4- to 10-fold greater rate of evolution has been consistently reported. The differences among studies are largely accounted for by differences in assumed divergence times. In contrast, the relative rate approach indicates an approximately 40% greater rate at nonsynonymous sites and no significant difference in rate for silent DNA including synonymous sites, 4-fold degenerate sites and noncoding sequences.

SLIGHTLY DELETERIOUS MUTATIONS REVISITED

In chapter 4 we discussed how the theory that molecular evolution involved the substitution of slightly deleterious mutations was introduced to explain a perceived discrepancy between the results of protein and DNA comparisons. This discrepancy is probably the consequence of inadequate methods of analysis, however, as we have already pointed out, this does not necessarily rule out the importance of slightly deleterious mutations in molecular evolution.

Table 6.1. Estimates of molecular evolutionary rate differences between primates and rodents

Molecular data	Rodent rate/Primate rate	Reference
Fossil-based estimates		
DNA-DNA hybridization	10	34
Amino acid sequence	~5	35
DNA-DNA hybridization	10	5
Amino acid sequence	5	36
DNA-DNA hybrid. + silent DNA	~5	4
Nucleotide sequence	4-10	7
Nucleotide sequence	4-8	8
DNA-DNA hybridization	~10	6
Synonymous sites	4.5	10
Relative-rate estimates		
Nonsynonymous sites	1.8	28
Nonsynonymous sites	1.5	29
Nonsynonymous sites	1.4	32
Nondegenerate sites	1.4	33
Nucleotide sequence	~1	27
Silent DNA	1.1	28
Silent DNA	1.1	33

The issue has continued to be investigated. Easteal and Collet's[33] comparison of 14 genes in marsupials, rodents and primates is mentioned above. They found a 40% greater rate of nondegenerate substitutions in rodents compared with primates, despite rate uniformity at noncoding and 4-fold degenerate sites. The increased nondegenerate substitution rate in rodents was seen at all but one of the fourteen genes and they concluded that this is best explained by invoking the substitution of slightly deleterious mutations.

In a recent analysis Ohta[37] compared the sequences of 17 genes in rodents, primates and artiodactyls. She came to what appeared to be an opposite conclusion, i.e. that, relative to the synonymous substitution rate, the nonsynonymous rate was slower in rodents than in primates. She also, by assuming a star phylogeny, inferred a greater rate of synonymous substitutions in the rodent lineage indicating a generation time effect. In fact the greater substitution rate in rodents represents the earlier divergence of rodents and thus the greater length of their lineage.

Ohta finds evidence in her analysis to support her nearly neutral model of molecular evolution. The evidence is that, relative to the synonymous substitution rate, the nonsynonymous rate in the rodent lineage has been slower that in the primate lineage, which is consistent with the earlier results of Gillespie.[38] If the population size in the rodent lineage has been larger than in the primate lineage, this difference is rate is expected for the substitution of slightly deleterious mutations.

The problem is that the rodent lineage, as defined by Ohta, is in fact only partly a rodent lineage. A substantial component of it is actually part of the branch leading to primates and artiodactyls following their separation from rodents. This point can be illustrated by reference to Figure 6.8 which shows the branching order of rodents, artiodactyls and primates relative to marsupials (the closest definite outgroup). This rooted tree shows that in addition to the terminal branches leading to the three placental orders (A, P, R) there is an internal branch that connects the node separating rodents from the other two orders with the node separating primates and artiodactyls. This branch is clearly not part of the rodent lineage, yet in Ohta's star phylogeny it is included in the rodent lineage. The consequence is that any difference observed between the "rodent" branch and the other two in the star phylogeny may be due to a difference in either the true rodent branch or in the internal branch or both. It is not necessarily a reflection of a difference in the real rodent lineage.

Easteal and Collet's[33] comparison of seven genes among marsupials, rodents, artiodactyls and primates shows that the synonymous/nonsynonymous ratio was approximately the same in the rodent and primate lineages but was substantially lower in the internal branch of

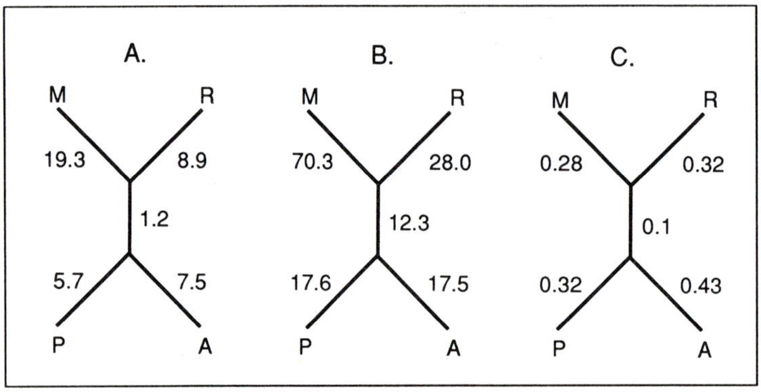

Fig. 6.8. Neighbor-joining trees connecting marsupials (M), rodents (R), artiodactyls (A) and primates (P), based on nucleotide substitution distances for nondegenerate (A) and 4-fold degenerate (B) sites, respectively, at seven genes. The number of substitutions per 100 bases on each branch is shown. (C) Ratios of nondegenerate to 4-fold degenerate substitution rates on each of the branches of the tree. Modified and reproduced with permission from S Easteal, C Collet, Mol Biol Evol 1994; 11:643-647.

the tree (Fig. 6.8). If a similar pattern exists for the genes analyzed by Ohta, the difference she reported would not be a result of any difference between rodents and primates or artiodactyls, but between the terminal branches leading to rodents, primates and artiodactyl on the one hand, and the internal branch on the primate/artiodactyl lineage on the other. A nearly neutral explanation of this difference would require that the population size of the common ancestor of primates and artiodactyls was greater than that on any of the terminal lineages leading to any of these three taxa.

EXAMPLES OF RATE VARIATION

We have made the case that in general the rate of evolution of homologous sequences is approximately the same in different mammalian lineages. We are not, however, implying that this is always the case. There are a number of well known exceptions and undoubtedly more will be discovered in time. When these exceptions occur as a result of the differential effect of natural selection they are of particular interest, because the provide an opportunity to investigate the process of adaptive evolution at the molecular level. A number of these cases have been reviewed by Gillespie.[22] These include: baboon hemoglobin, visual pigment genes and human cytochromes. Other examples include rodent insulins,[39] the stomach lysozymes of foregut fermenters[40] and growth hormone genes.[41]

SUMMARY

Differences in evolutionary rate as great as 10-fold have been identified between rodents and primates, but only by methods that depend on an interpretation of the fossil record to derive divergence times. In contrast, when a valid phylogeny is used and the relative approach to testing for evolutionary rate variation is applied, no rate difference is evident in noncoding DNA. A discrepancy of exactly this kind was previously found in the analysis of primates (discussed in chapter 5), where it resulted in a substantial reinterpretation of the primate fossil record. The discrepancy between the two approaches in the comparison of primates and rodents requires a similar reassessment of the mammalian fossil record. We discuss this and related issues in detail in the next four chapters.

REFERENCES

1. Easteal S. Generation time and the rate of molecular evolution. Mol Biol Evol 1985; 2:450-453.
2. Li W-H, Wu C-I. Rates of nucleotide substitution are evidently higher in rodents than in man. Mol Biol Evol 1987; 4:74-77.
3. Easteal S. The rates of nucleotide substitution in the human and rodent lineages: a reply to Li and Wu. Mol Biol Evol 1987; 4:78-80.
4. Britten RJ. Rates of DNA sequence evolution differ between taxonomic groups. Science 1986; 231:1393-1398.
5. Brownell E. DNA/DNA hybridization studies of muroid rodents: symmetry and rates of molecular evolution. Evolution 1983; 37:1034-1051.
6. Catzeflis F, Sheldon FH, Ahlquist JE, Sibley CG. DNA-DNA hybridization evidence of the rapid rate of muroid rodent DNA evolution. Mol Biol Evol 1987; 4:242-253.
7. Li W-H, Tanimura M. The molecular clock runs more slowly in man than in apes and monkeys. Nature 1987; 326:93-96.
8. Li W-H, Tanimura M, Sharp PM. An evaluation of the molecular clock hypothesis using mammalian DNA sequences. J Mol Evol 1987; 25:330-342.
9. Holmes EC. Different rates of substitution may produce different phylogenies of the eutherian mammals. J Mol Evol 1991; 33:209-215.
10. Bulmer M, Wolfe KH, Sharp PM. Synonymous nucleotide substitution rates in mammalian genes: implications for the molecular clock and the relationship of mammalian orders. Proc Natl Acad Sci (USA) 1991; 88:5974-5978.
11. O'hUigin C, Li W-H. The molecular clock ticks regularly in muroid rodents and hamsters. J Mol Evol 1992; 35:377-384.
12. Wilson AC, Ochman H, Prager EM. Molecular time scale for evolution. Trends Genet 1987; 3:241-247.
13. Chevret P, Denys C, Jaeger J-J, Michaux J, Catzeflis FM. Molecular evidence that the spiny mouse (*Acomys*) is more closely related to gerbils

(Gerbillinae) than to true mice (Murinae). Proc Natl Acad Sci (USA) 1993; 90:3433-3436.
14. Ohta T, Kimura M. On the constancy of the evolutionary rate of cistrons. J Mol Evol 1971; 1:18-25.
15. Romer AS. Vertebrate paleontology, 3rd edition. Chicago: University of Chicago Press, 1966.
16. Carroll RL. Vertebrate paleontology and evolution. New York: W.H. Freeman & Co, 1988.
17. Townes TM, Shapiro SG, Wenke SM, Lingrel JB. Duplication of a four-gene set during the evolution of the goat β-globin locus produced genes now expressed differentially in development. J Biol Chem 1984; 259:1896-1900.
18. Langley CH, Fitch WM. An examination of the constancy of the rate of molecular evolution. J Mol Evol 1974; 3:161-177.
19. Fitch WM, Langley CH. Protein evolution and the molecular clock. Fed Proc 1976; 35:2092-2097.
20. Kimura M. The neutral theory of molecular evolution. Cambridge: Cambridge University Press, 1983.
21. Takahata N. On the overdispersed molecular clock. Genetics 1987; 116:169-179.
22. Gillespie JH. The molecular clock may be an episodic clock. Proc Natl Acad Sci (USA) 1984; 81:8009-8013.
23. Gillespie JH. Natural selection and the molecular clock. Mol Biol Evol 1986; 3:138-155.
24. Gillespie JH. Variability of evolutionary rates of DNA. Genetics 1986; 113:1077-1091.
25. Gillespie J. The causes of molecular evolution. Oxford: Oxford University Press, 1991.
26. Wu C-I, Li W-H. Evidence for higher rates of nucleotide substitution in rodents than in man. Proc Natl Acad Sci (USA) 1985; 82:1741-1745.
27. Easteal S. Rate constancy of globin gene evolution in placental mammals. Proc Natl Acad Sci (USA) 1988; 85:7622-7626.
28. Easteal S. The pattern of mammalian evolution and the relative rate of molecular evolution. Genetics 1990; 124:165-173.
29. Li W-H, Gouy M, Sharp PM, O'hUigin C, Yang Y-W. Molecular phylogeny of Rodentia, Lagomorpha, Primates, Artiodactyla, and Carnivora and molecular clocks. Proc Natl Acad Sci (USA) 1990; 87:6703-6707.
30. Janke A, Feldmaier-Fuchs G, Thomas WK, Von Haeseler A, Pääbo S. The marsupial mitochondrial genome and the evolution of placental mammals. Genetics 1994; 137:243-256.
31. Graur D. Molecular phylogeny and the higher classification of eutherian mammals. Trends Ecol Evol 1993; 8:141-147.
32. Gu X, Li W-H. Higher rates of amino acid substitution in rodents than in humans. Mol Phyl Evol 1992; 1:211-214.
33. Easteal S, Collet C. Consistent variation in amino-acid substitution rate, despite uniformity of mutation rate: protein evolution in mammals is not

neutral. Mol Biol Evol 1994; 11:643-647.
34. Laird CD, McConaughy BL, McCarthy BJ. Rate of fixation of nucleotide substitutions in evolution. Nature 1969; 224:149-154.
35. Goodman M, Braunitzer G, Stangl A, Schrank B. Evidence on human origins from haemoglobins of African apes. Nature 1983; 303:546-548.
36. Goodman M. Rates of molecular evolution: the hominoid slowdown. BioEssays 1985; 3:9-14.
37. Ohta T. An examination of the generation-time effect on molecular evolution. Proc Natl Acad Sci (USA) 1993; 90:10676-10680.
38. Gillespie JH. Lineage effects and the index of dispersion of molecular evolution. Mol Biol Evol 1989; 6:636-647.
39. Beintema JJ, Campagne RN. Molecular evolution of rodent insulins. Mol Biol Evol 1987; 4:10-18.
40. Stewart CB, Schillig JW, Wilson AC. Adaptive evolution in the stomach lysozymes of foregut fermenters. Nature 1987; 330:401-404.
41. Wallis M. Variable evolutionary rates in the molecular evolution of mammalian growth hormones. J Mol Evol 1994; 38:619-627.

CHAPTER 7

The Mammalian Fossil Record

We have argued in earlier chapters that rates of DNA evolution appear to be generally uniform among mammalian orders. We have also suggested that, in many studies, the lack of apparent rate uniformity has been based on inappropriate assumptions about species' divergence times. We have implied that at least some of those assumed divergence times of mammals need to be reassessed. In chapter 5, we discussed the most celebrated case in which a divergence time was reassessed in the light of molecular data. This involved the split between humans and other African apes. To discuss the implications of the molecular clock in more detail we need to briefly document the mammalian fossil record with special reference to events that are important in relation to the molecular clock debate. In this chapter we discuss fossils relevant to the separation of marsupials and placentals, the radiation of the placental mammalian orders, and the divergence times of murine rodents. In the next chapter we focus on the fossil evidence for primate divergence times.

Our argument for a molecular clock is based on relative rate analyses that do not depend on fossil-based divergence times. The demonstration of rate constancy in this way can be used to reassess the fossil evidence. We will discuss the divergence time implications of a molecular clock in more detail in chapter 10. Briefly, the separation of placentals from marsupials, of placental orders from each other, and of mice from rats must have occurred much earlier than is currently believed, and many of the primate divergences must have occurred much later. Here and in the next chapter, we examine the relevant fossil evidence to provide a basis for evaluation of existing theories about mammalian divergence times, and to determine if there is sufficient evidence to rule out the early or late divergences predicted by molecular data. In chapter 9 we will extend this analysis to consider the distributions of fossil and extant species in the context of what is known of continental drift. The discussion in the next three chapters requires knowledge of the geologic time scale. Readers unfamiliar with this should refer to Table 1.1.

THE MAMMALIAN FOSSIL RECORD

Most mammalian orders are thought to have originated in Europe and North America. This is because many of their earliest known fossils (from the Paleocene and late Cretaceous) are from these continents and because paleontologists tend to follow "Matthew's Rule", that taxa probably evolved where their oldest representatives are found.[1] It should be realized, however, that of all continents, Europe and North America have the highest concentration of paleontologists, and that the finding there of early mammal fossils may reflect a bias in sampling and a comparative lack of evidence from the southern continents. The earliest fossil of a group provides evidence only that the group was in existence by the time the fossil was deposited, and much of the world has yet to be systematically searched for fossils.[2]

Many mammalian fossils are fragmentary, comprising teeth or cranial fragments such as pieces of dentaries. This is markedly so in the sparse Mesozoic mammalian record.[3-5] The systematics of extant mammals is also based on post-cranial features and soft-tissue characters which are rarely if ever preserved in known Mesozoic mammal fossils, which complicates their interpretation. Conversely, extant taxa cannot all be distinguished by skeletal remains, so related paleospecies may not be recognized as distinct even if their skeletons are fully preserved.

Mesozoic placental mammals were not recognized as such until the 1920s, and are still poorly sampled in comparison to Tertiary fossils. Improving knowledge of this fauna, particularly from late Cretaceous North America and Asia, reveals considerable placental diversity before the Cretaceous-Tertiary mass extinction.[6] However, much of the world has been poorly sampled. For instance, until recently only one really good site was known in Africa (Fayum, Egypt) covering the period from the Paleocene to the Oligocene. Before the Paleocene, there is a 120 Ma gap with very few known African mammals. Yet, as suggested by present-day distributions, Africa may well have been central to the evolution of several mammalian orders.

WHAT FOSSILS ARE MAMMALS?

Discussion of mammalian origins is complicated by disagreement over what fossils should formally be called mammals. Traditionally, "true mammals" were arbitrarily defined by possession of a dentary-squamosal jaw articulation, shared by no other animals. This trait first appears in fossils from the Triassic-Jurassic boundary (about 210 Ma ago). However, other obvious diagnostic features of living mammals may have appeared even earlier. Fur may have developed by the early Triassic, associated with true endothermy, and homeothermy may have preceded the dentary-squamosal jaw articulation by as much as 40 Ma.[7-10]

Cladistic approaches to taxonomy[11] have resulted in phylogenetic definitions of Mammalia. These are explicitly based and tested on evident shared descent from a putative common ancestor. Although we are interested in phylogeny, we are not concerned here with cladistic formalism. For those that are, however, we need to make clear what we are talking about. In referring to mammals we use a "stem-group" definition for Mammalia: "the *earliest* common ancestor of living mammals that was also *not an ancestor of other living amniotes* (birds, crocodilians, snakes and lizards, turtles), and all its descendants that have ever lived".[12,13] The alternative is a "crown-group" definition: "the *last* common ancestor of all living mammals, and all its descendants that have ever lived".[14] The latter definition excludes the so-called mammal-like reptiles.

THE ORIGIN OF MAMMALS

The mammalian fossil record is the longest and most continuous of any extant amniote class.[4,7,16-17] The mammalian stem group goes back to the Synapsida, which appear in the fossil record in the latter half of the Carboniferous period, about 290 and possibly 320 Ma ago. These dates indicate the latest possible separation time between mammals and all the other living amniotes.

From 280 Ma ago, there is a well-documented succession of nested synapsid groups, such as the therapsids and then the cynodonts, which progressively acquired skeletal hallmarks of Mammalia over a 100 Ma period. The earliest "true mammals" (dentary-squamosal jaw joint) are known from the late Triassic to early Jurassic of Europe, Asia and Africa. Some of these taxa have postcranial remains that are similar to those of modern, small, foraging insectivores.[18] The mammalian fossil record is extremely sparse over the next 140 Ma, and mostly comprises teeth. Not until the latest Cretaceous (70 Ma ago), are there fossil taxa that may belong to modern placental orders.[4,10,14,16,17,19]

THE SPLIT BETWEEN THERIANS (MARSUPIALS AND PLACENTALS) AND MONOTREMES

A well-established *minimum* date for the monotreme-therian divergence would place a *maximum* limit on the divergence of marsupials and placentals. The latter are clearly each other's sister groups, and together are often described as Theria.[14]

Monotremes are the only survivors of a large Mesozoic array of nontribosphenid mammals (see below). Suggested monotreme origins have been as early as the lower mid-Jurassic, based on reptilian traits such as the monotreme pectoral girdle.[20] Romer[7] offered as potential ancestors either symmetrodonts (Late Triassic-Early Jurassic) or mid-Jurassic pantotheres. Rowe[14] gives an early-Jurassic date for the split, based on the appearance of *Morganucodon* as a sister-group to all living mammals. In this view, the monotreme-therian split could be at least as old as the late Triassic (>200 Ma ago). Others suggest monotremes appeared later and closer to modern therians, possibly diverging as late as the Late Cretaceous, although both monotreme[21,22] and therian fossils are now known from the early Cretaceous. On the present scanty evidence, estimating when in the preceding 150 Ma of mammalian evolution the therians and monotremes diverged is little more than guesswork.

THE EARLIEST MARSUPIALS AND PLACENTALS

Extant placentals and marsupials are readily distinguished by, among other things, dental and jaw traits. The distinction in Mesozoic fossils, however, is not as simple. Some dental and jaw characteristics of extant marsupials are controversial and may be ancestral for all therians. Also, placentals and marsupials represent just two surviving therian lineages from a once more diverse array. Therian fossils that cannot be assigned to either group overlap the earliest definite placentals and marsupials, some persisting as late as the Cretaceous-Tertiary mass extinction.[9,19,23]

All living therians are tribosphenids, which are distinguished by the presence of a particular type of molar, though in some descendant taxa this trait has secondarily been lost. Unequivocal therians are first identified by three molars from early Cretaceous Europe, Asia and North America. None of these fossils can be identified definitely either as marsupial or placental. The oldest fully accepted placentals are from late Cretaceous Mongolia. There is, however, some inadequately described material from Central Asia that may be late Early Cretaceous.[4,7,24,25]

Fossil marsupials are known from all continents, including Antarctica. Until recently, the oldest known marsupials were represented by a maxillary fragment and teeth from mid-Cretaceous beds (100 Ma) in Texas and Utah respectively.[14,15,26] This implied a minimum date for the split of 95-100 Ma. However, there is recent evidence from southwestern Utah of marsupials possibly as early as 140 Ma ago.[25] Furthermore, Marshall and Kielen-Jaworowska[27] have recently suggested that tribosphenic molars were acquired independently by marsupials and placentals. If this was the case the separation of the two groups may have occurred well before the first occurrence of this form of dentition in the Early Cretaceous.

Thus on fossil evidence, the two lineages could well have been distinct at least since the early Cretaceous or late Jurassic, but the location in time and space of their last common ancestor remains unknown.[24] An early Jurassic, or even a Triassic divergence cannot be ruled out by the fossil evidence.

THE ORIGIN OF PLACENTAL ORDERS

A widely held view of the phylogeny of placental orders is that they diverged in a star-like radiation near the time of the Cretaceous-

Tertiary mass extinction. This well-known event is thought to have paved the way for the "age of mammals". The mass extinction occurred 65 Ma ago and is now thought to have been caused by the impact of a 6-14 km diameter asteroid.[28]

The impact was first detected indirectly by the occurrence of an extremely narrow but marked peak of Platinum-group elements (particularly Iridium) in contemporaneous sediments. This anomaly is now known to occur worldwide, and the isotopic composition of the sediments and of associated mineral-glass "tektites"[29,30] are thought to be inconsistent with volcanic origins. There is now excellent evidence of an enormous impact crater of precisely the right age near Chicxulub in Mexico, on the northern shore of the Yucatan Peninsula.[31,32] Though initially controversial, this impact hypothesis has now become widely accepted,[33,34] and the risk of such impacts has become a matter of serious study.[35]

We saw in chapter 6 how molecular evidence has shown that there was a dichotomous branching rather than a star-like radiation of at least some placental orders. The question is whether these branching events occurred before or after the Cretaceous-Tertiary mass extinction. Are divergence times in the Mesozoic at least consistent with the fossil evidence?

Late Cretaceous placentals are well known in the northern continents, though most remains are fragmentary. The earliest alleged placentals, as mentioned above, come from Mesozoic Central Asia, possibly of late Early Cretaceous age, but have yet to be fully described.[24] In North America, where the Mesozoic mammal record is better known, there are mammal fossils in the 85-100 Ma period.[36] However, these cannot for the most part be identified as belonging to specific modern orders.

The earliest known fossils of almost all modern placental orders are in the Paleogene, after the Cretaceous-Tertiary mass extinction. Exceptions include some late Cretaceous fossils that are assigned to the ungulate "condylarths", the alleged primate affinities of a single Cretaceous molar tooth (*Purgatorius*), and the upper Cretaceous *Batodon* from North America,[4,37] a probable insectivore. Certain times and places for first known appearances of placental orders can be established from the fossil record.[3,4,7,8,38-59] There is little fossil evidence of modern placental orders being established before the end of the Cretaceous. However, it would be wrong to conclude that placental mammals did not diverge during the Cretaceous. The therian fossil record during the whole of the Cretaceous (144-65 Ma ago) is too poor to allow definite conclusions to be made about dates of origins of specific orders. The prevailing view of a radiation at or soon after the Cretaceous-Tertiary mass extinction derives more from ignorance of what happened during the Cretaceous than from knowledge of what happened during the Tertiary.

The derived traits that characterize modern placental orders tend to be associated with ecological diversity. The Cretaceous fossil record may yet reveal greater diversity of placental forms. However, even if it does not, it remains entirely possible that the placental orders were well established before the end of the Cretaceous, but that they had not, by then, diversified ecologically and were therefore morphologically similar. The Cretaceous-Tertiary mass extinction may have provided the opportunity for ecological, and hence morphological, diversification of lineages that had already been distinct for a long time.[60]

THE SEPARATION OF *MUS* AND *RATTUS*

Estimates of the separation time of mice and rats (Murinae) have been critical to debate on the molecular clock, leading some to conclude that rates of molecular evolution are faster in rodents than in primates. In fact, the fossil record of murid rodents is controversial.

Progonomys is the earliest definite murine, and is known from 12 Ma ago at scattered sites in Ethiopia, China and Europe. This dispersed genus has been suggested as ancestral to both *Mus* and *Rattus*, giving a maximum split of 12 Ma ago. Jaeger et al,[61] however, have suggested that *Progonomys* was ancestral only to some species in *Mus*. This would imply a *Mus-Rattus* split much earlier than 12 Ma ago.

Antemus from the Siwaliks, Pakistan, variously dated to 14 Ma or to 16-17 Ma ago, is the earliest definite murid. It is an alleged sister group both to living murines (including mice and rats) and perhaps to *Progonomys*. Similar fossils ("primitive dendromurids") of slightly greater age are known from Asia, and from three sites in Africa.[61,62] If South Asian *Antemus* really was a sister group to extant murines, the split between *Rattus* and *Mus* should be later than its first appearance (17 Ma ago). However, the first appearance of *Antemus* could be misleading if, say, it were a surviving relic or if murines arose elsewhere from other earlier dendromurids. The murid fossil record is meager and there is little consensus on its interpretation. In chapter 6 we discussed the implications for the rat-mouse divergence time of recent molecular studies on the phylogeny of Muridae. A late divergence (12 Ma) is consistent with the fossil record, but an earlier divergence cannot be ruled out, and is implicated by the molecular phylogeny.

The divergence time between rats and mice must postdate the divergence of the Muridae from their presumed sister group the Cricetidae. It had been suggested that cricetids are ancestral to murids, and thus that the murid-cricetid divergence occurred after the late Eocene, which is the date of the oldest known cricetid fossil. However, morphological convergence and parallelism are widespread in rodent evolution[63] and this interpretation may not be correct. The murid-cricretid divergence may have been during the Eocene or earlier if it occurred before the first cricetid fossil. This would allow an Eocene or earlier divergence of mice from rats.

SUMMARY

Mammals appear to have diverged from other amniotes more than 300 Ma ago. The relevant fossil record during the next 200 Ma is too poor to allow any real estimate of the divergence of monotremes from therians. Because of this, marsupials and placentals may have diverged at any time between approximately 150 and 250 Ma ago. There is some fragmentary, but no compelling, evidence that at least some orders of placental mammals diverged before the Cretaceous/Tertiary boundary, 65 Ma ago. The mammalian fossil record during the Cretaceous, however, is so poor that no definite conclusions about divergence times are possible. The same is true of the rodent fossil record which allows a 50 Ma ago, or earlier, divergence of mice and rats.

We conclude that current estimates of the divergence times we have discussed are not excluded by the fossil data, but that neither are much earlier divergence times. In the context of evaluating the molecular clock, *actual* and not *minimum* divergence times are important. These should not be confused; they may be very different when the fossil record is poor.

References

1. Kirsch J. Vicariance biogeography. In: Archer M, Clayton G, eds. Vertebrate zoogeography and evolution in Australasia. Western Australia: Hesperian Press, 1984:109-112.
2. Martin RD. Primate origins and evolution, a phylogenetic reconstruction. London: Chapman and Hall, 1990.
3. Kielan-Jaworowska Z, Crompton AW, Jenkins FA. The origin of egg-laying mammals. Nature 1987; 326:871-873.
4. Carroll RL. Vertebrate paleontology and evolution. New York: W.H. Freeman, 1988.
5. Szalay FS, Novacek MJ, McKenna MC. Mammal phylogeny. New York: Springer-Verlag, 1993.
6. Cifelli RL. A primitive higher mammal from the late Cretaceous of Southwestern Utah. J Mammal 1990; 71:343-350.
7. Romer AS. Vertebrate paleontology. Chicago: University of Chicago Press, 1966.
8. McNab BK. The evolution of endothermy in the phylogeny of mammals. Am Nat 1978; 112:1-21.
9. Archer M. Origins and early radiations of mammals. In: Archer M, Clayton G, eds. Vertebrate zoogeography and evolution in Australasia. Western Australia: Hesperian Press, 1984:477-516.
10. Benton MJ. Phylogeny of the major tetrapod groups: morphological data and divergence dates. J Mol Evol 1990; 30:409-424.
11. Hennig W. Phylogenetic systematics. Urbana: University of Ilinois Press. 1966.
12. Gee H. By their teeth ye shall know them. Nature 1992; 360:529.
13. Forey PL. Therapsids and transformation series. Nature 1993; 361:596-597.

14. Rowe T. Phylogenetic systems and the early history of mammals. In: Szalay FS, ed. Mammal phylogeny. New York: Springer-Verlag, 1993:129-145.
15. Archer M. Origins and early radiations of marsupials. In: Archer M, Clayton G, eds. Vertebrate zoogeography and evolution in Australasia. Western Australia: Hesperian Press, 1984:585-626.
16. Kemp TS. A note on the Mesozoic mammals, and the origins of therians. In: Benton MJ, ed. The phylogeny and classification of the tetrapods, volume 2: Mammals. Oxford: Clarendon, 1988:23-29.
17. Laurin M, Reisz RR. *Tetraceratops* is the oldest known therapsid. Nature 1990; 345:249-250.
18. Jenkins FA, Parrington FR. The postcranial skeletons of the Triassic mammals *Eozostrodon, Megazostrodon* and *Erythrotherium.* Phil Trans R Soc London 1976; 273:3867-4300.
19. Lillegraven JA, Kielan-Jaworowska Z, Clemens WA. Mesozoic mammals: the first two thirds of mammalian history. Berkeley: University of California Press, 1979.
20. Murray P. Furry egg-layers: the monotreme radiation. In: Archer M, Clayton G, eds. Vertebrate zoogeography and evolution in Australasia. Western Australia: Hesperian Press, 1984:571-584.
21. Archer MA, Flannery TF, Ritchie A, Molnar RE. Mesozoic mammal from Australia—an early Cretaceous monotreme. Nature 1985; 363-366.
22. Archer M, Murray P, Hand S, Godthelp H. Reconsideration of monotreme relationships based on the skull and dentition of the Miocene *Obdurosdon dicksoni.* In: Szalay FS, ed. Mammal phylogeny. New York: Springer-Verlag, 1993:75-94.
23. Ciffeli RL. Theria of metatherian-eutherian grade and the origin of marsupials. In: Szalay FS, ed. Mammal phylogeny. New York: Springer-Verlag, 1993:205-215.
24. Lillegraven JA, Thompson SD, McNab BK, Patton JL. The origin of eutherian mammals. Biol J Linn Soc 1987; 32:281-336.
25. Eaton JG. Therian mammals from the cenomanian (Upper Cretaceous) Dakota formation, southwestern Utah. J Vert Paleontol 1993; 13:105-124.
26. Aplin KP, Archer M. Recent advances in marsupial systematics with a new syncrestic classification. In: Archer M, ed. Possums and opossums: studies in evolution. Chipping Norton, New South Wales: Surrey Beatty & Sons, 1987:xv-lxxii.
27. Marshall LG, Kielan-Jaworowska Z. Relationships of the dog-like marsupials, deltatheroidans and early tribosphenic mammals. Lethaia 1992; 25:361-374.
28. Alvarez LW, Alveraz W, Asaro F, Michel HV. Extraterrestrial cause for the Creaceous-Tertiary extinction. Science 1980; 208:1095-1108.
29. Sigurdsson H, Bonte P, Turpin L, Chaussidon M, Metrich N, et al. Geochemical constraints on source region of Cretaceous/Tertiary impact glasses. Nature 1991; 353:839-842.
30. Sigurdsson H, D'Hondt S, Arthur MA, Bralower TJ, Zachos JC, et al. Glass from the Cretaceous/Tertiary boundary in Haiti. Nature 1991;

349:482-487.
31. Krogh TE, Kamo SL, Sharpton VL, Marin LE, Hildebrand AR. U-Pb ages of single shocked zircons linking distal K/T ejecta to the Chicxulub crater. Nature 1993; 366:731-734.
32. Sharpton VL, Burke K, Camargo-Zanoguera S, Hall SH, Lee DS, et al. Chicxulub multiring impact basin: size and other characteristics derived from gravity analysis. Science 1993; 261:1564-1567.
33. Kerr RA. Yucatan killer impact gaining support. Science 1991; 252:377.
34. Kerr RA. Testing an ancient impact's punch. Science 1994; 263:1371-1372.
35. Chapman CR, Morrison D. Impacts on the Earth by asteroids and comets: assessing the hazards. Nature 1994; 367:33-39.
36. Marshall CR. The fossil record and estimating divergence times between lineages: maximum divergence times and the importance of reliable phylogenies. J Mol Evol 1990; 30:400-408.
37. Kielan-Jaworowska Z, Bown TM, Lillegraven JA. Eutheria. In: Lillegraven JA, Kielan-Jaworowska Z, Clemens WA, eds. Mesozoic mammals: the first two thirds of mammalian history. Berkeley: University of California Press, 1979:221-258.
38. Simpson GG. The principles of classification and a classification of mammals. Bull Am Mus Nat Hist 1945; 85:1-350.
39. Gentry AW, Hooker JJ. The phylogeny of the Artiodactyla. In: Benton MJ, ed. The phylogeny and classification of the tetrapods, volume 2: Mammals. Oxford: Clarendon Press, 1988:235-271.
40. Martin LD. Fossil history of the terrestrial Carnivora. In: Gittleman JL, ed. Carnivore behaviour and evolution. London: Chapman & Hall, 1989:536-568.
41. Gingerich PD, Wells NA, Russell DA, Ibrahim-Shah SM. Origin of whales in epicontinental remnant seas: new evidence from the early Eocene of Pakistan. Science 1983; 220:403-406.
42. Gingerich PD, Smith BH, Simons EL. Hind limbs of Eocene *Basilosaurus:* evidence of feet in whales. Science 1990; 249:154-157.
43. Thewissen JGM, Hussain ST, Arif M. Fossil evidence for the origin of aquatic locomotion in archaeocete whales. Science 1993; 263:210-212.
44. Hand SJ, Novacek MJ, Archer M. An early Tertiary bat from the Tingamarra local fauna of southeastern Queensland, paper & abstract. In: Fourth Australian Bat Conference, University of Queensland, 1991.
45. Smith JD. Comments on flight and the evolution of bats. In: Hecht MK, Goody PC, Hecht BM, eds. Major patterns in vertebrate evolution. New York: Plenum Press, 1977:427-437.
46. Hand S. Bat beginnings and biogeography: a southern perspective. In: Archer M, Clayton G, eds. Vertebrate zoogeography and evolution in Australasia. Western Australia : Hesperian Press, 1984:853-905.
47. Beard KC. Gliding behaviour and palaeoecology of the alleged primate family Paromomyidae (Mammalia, Dermoptera). Nature 1990; 345:340-341.

48. Kay RF, Thorington RW, Houde P. Eocene plesiadapiform shows affinities with flying lemurs, not primates. Nature 1990; 345:342-344.
49. Krause DW. Were paromomyids gliders? Maybe, maybe not. Hum Evol 1991; 21:177-188.
50. Li C-K, Ting S-Y. Possible phylogenetic relationships of Asiatic eurymylids and rodents, with comments on mimotonids. In: Luckett WP, Hartenberger J-L, eds. New York: Plenum Press, 1985:35-58.
51. Butler PM. Phylogeny of the insectivores. In: Benton MJ, ed. The phylogeny and classification of the tetrapods. Oxford: Clarendon Press, 1988:117-141.
52. McKenna MC, Minchen C, Suyin T, Zhexi L. *Radinskya yupingae*, a perissodactyl-like mammal from the late Paleocene of southern China. In: Prothero DR, Schoch RM, eds. The evolution of perissodactyls. Oxford: Clarendon Press, 1989:24-36.
53. Novacek MJ. Fossils, topologies, missing data, and the higher level phylogeny of eutherian mammals. Syst Biol 1992; 41:58-73.
54. Thewissen JGM, Domning DP. The role of phenacodontids in the origin of the modern orders of ungulate mammals. J Vert Paleontol 1992; 12:494-504.
55. Storch G. *Eomanis waldi*, ein Schuppentier aus dem Mittel-Eozän der "Grube Messel" bei Darmstadt (Mammalia: Pholidota) Fossilfundstelle Messel. Senckenbergiana Lethaea 1978; 59:503-529.
56. Court N. Cochlea anatomy of *Numidotherium koholense*: auditory acuity in the oldest known proboscidean. Lethaia 1992; 25:211-215.
57. Jaeger J-J, Denys C, Coiffat B. New Phiomorpha and Anomaluridae from the late Eocene of Northwest Africa: phylogenetic implications. In: Luckett WP, Hartenberger J-L, eds. Evolutionary relationships among rodents: a multidisciplinary analysis. New York: Plenum, 1985:567-588.
58. Luckett WP, Hartenberger J-L. Evolutionary relationships among rodents: comments and conclusions. In: Luckett WP, Hartenberger J-L, eds. Evolutionary relationships among rodents: a multidisciplinary analysis. New York: Plenum, 1985:685-712.
59. Storch G. *Eirptamandua joresi*, ein Myrmecophagide aus dem Eozän der "Grube Messel" bei Darmstadt (Mammalia: Xenarthra). Fossilfundstelle Messel Nr 19. Senckenbergiana Lethaea 1981; 61:247-289.
60. Jablonski D. Evolutionary consequences of mass extinctions. In: Raup DM, Jablonski D, eds. Patterns and processes in the history of life. Berlin: Springer-Verlag, 1986:313-329.
61. Jaeger J-J, Tong H, Denys C. The age of the *Mus-Rattus* divergence: paleontological data compared with the molecular clock. C R Acad Sc Paris 1986; 302:917-922.
62. Flynn LA, Jacobs LL, Lindsay EH. Problems in muroid phylogeny: relationships to other rodents and origin of major groups. In: Luckett WP, Hartenberger J-L, eds. Evolutionary relationships among rodents: a multidisciplinary analysis. New York: Plenum Press, 1985:589-616.
63. Luckett WP, Hartenberger J-L, eds. Evolutionary relationships among rodents: a multidisciplinary analysis; New York, Plenum Press 1985.

CHAPTER 8

Primate Separation Times from the Fossil Record

The divergence times of certain primate taxa are important to the molecular clock debate. We have already discussed in chapter 5 how molecular data had a profound effect on our understanding of the relationships among humans and other apes. In this chapter we present an abbreviated account of the enormous literature that encompasses the interpretation of fossil primates. We focus on those fossils that are important for the interpretation of molecular data. Estimates of minimum dates for the crucial splits, and the fossils from which they are derived, are listed in Table 8.1. We emphasize that all estimates of this kind are minimum estimates and that they are provisional. Our estimates are no exception. They may be revised either downwards or upwards by new fossil discoveries or by reinterpretation of known fossils.

A broad outline of primate phylogeny, with emphasis on the lineage leading to humans, was given in chapter 4. The primary division within primates was between strepsirhines (lemurs, lorises and galagos) and haplorhines (apes, monkeys and tarsiers). The first division within haplorhines was between tarsiers and simians (apes and monkeys). Within the simians the platyrrhines (New World Monkeys) diverged from the catarrhines (Old World monkeys and apes), followed by the separation of cercopithecoids (Old World monkeys) from hominoids (apes). The first divergence within the apes was between hylobatids (gibbons and siamangs) and hominids (great apes). Within the great apes, the Asian orangutan, *Pongo,* diverged first from the African apes, including humans. The divergence order of the African apes is still contentious, but molecular data now strongly support a divergence of *Gorilla* before the separation of *Pan* (chimpanzee) and *Homo.*

EARLY FOSSILS

Relationships between primates and other placental orders are uncertain and the fossil record is of little help. Primates were once thought to belong to the superorder Archonta which also included tree shrews

(Scandentia), elephant shrews (Macroscelidea), flying lemurs (Dermoptera), and perhaps bats (Chiroptera). However, these relationships now have little support.[1-4]

The earliest alleged primate fossil is a single lower molar tooth attributed to *Purgatorius* from Late Cretaceous deposits at Hells Creek, Montana.[5] The more widespread mid-Paleocene *Palaechthon* is similar. If the diagnosis and systematics of *Purgatorius* are correct this gives a 67 Ma minimum age for the origin of primates. Both *Purgatorius* and *Palaechthon* belong to the Plesiadapiformes, a group of Cretaceous to Eocene mammals from Europe and North America. Although these have long been considered primates, this is now uncertain[4,6,7] and it has recently been suggested that plesiadapiforms have a closer affinity to dermopterans.[8-10] If plesiadapiforms are not primates then no fossil primates are known from the Mesozoic.[11-12]

Other fossils have been reported in the past decade that indicate a 56-60 Ma minimum date for primate origins. *Decoredon* (about 60 Ma) is a probable primate from mid Paleocene China.[13] *Altiatlasius*, a haplorhine from the Paleocene (about 57 Ma) and the slightly later *Biretia*, both from North Africa, may be simians (monkeys and apes).[14] If this is confirmed, it places the primary split within the haplorhines (i.e. between tarsiers and simians) in the early Paleocene at the latest. Therefore, the separation between haplorhines and strepsirhines would have been even earlier.

The earliest fossils possibly belonging to the strepsirhine lineage are the Adapiformes, an Eocene to late Miocene group that resemble lemurs. Eocene adapiforms are well known from North America and Europe and now from China.[15] The earliest known are *Cantius* and *Teilhardina* (very early Eocene) and then *Notharctus*.[16,17] Lemurs have a very poor fossil record. The oldest known loris and galago fossils (*Progalago*, *Komba* and *Mioeuoticus*) come from several early Miocene sites in East Africa.[16,18,19] These show that both groups have been established for at least 20 Ma.

Within the haplorhines, a minimum date for the simian/tarsier split is indicated by the early Eocene appearance of the first Omomyiformes, which are thought to include the tarsier lineage. This again implies earlier dates for the strepsirhine/haplorhine split and the origin of Primates. The earliest omomyiforms known from Europe are the early Eocene *Nannopithex* and the middle Eocene *Pseudoloris*.[19,20] In North America, *Shoshonius* is the earliest, dating to 50.5 Ma.[21] Two definite tarsiers (*Macrotarsius* and *Tarsius* itself) and a simian have recently been described from the mid Eocene of southeastern China.[15]

Within the simians, the split between the catarrhines (apes and Old World monkeys) and the platyrrhines (New World monkeys) may have a late Eocene minimum date, based on the first appearance of catarrhines. Two fossil simians from the late Eocene of Burma, *Pondaungia* and *Amphipithecus*, have been dated to 40-44 Ma ago.[22-24]

These share derived teeth and jaw characteristics with catarrhines.[11,25] Beard et al[15] have recently described the simian *Eosimias* from the mid Eocene (about 45 Ma) of southeastern China, which resembles *Amphipithecus*. Beard et al[15] consider that all these antedate the catarrhine/platyrrhine split. If, however, these Asian fossils turn out to be catarrhines, they would give a latest possible time for the catarrhine/platyrrhine split of about 45 Ma.

The oldest fossils widely accepted as catarrhine are from late Eocene Egypt (about 36 Ma). Deposits in the Fayum Depression contain many species, notably *Aegyptopithecus* and *Catopithecus* that are now thought to be late Eocene and not Oligocene as previously believed.[26-28] Older reports suggested that many Fayum fossils belonged to particular extant catarrhine families. This implied that the ape/Old World monkey split had already been established.[29-31] However, current opinion is that similarities of Fayum catarrhines to extant families are due to convergence and that all Fayum fossils so far discovered antedate the split between the hominoids (apes) and the cercopithecoids (Old World Monkeys).[19,26,32-35]

The oldest known platyrrhine (New World monkey), *Branisella boliviana*, comes from latest Oligocene deposits at Salla, Bolivia, and is aged about 26 Ma. It cannot be ascribed to any modern platyrrhine family.[19,36] The origin of the New World monkeys has long been controversial.[18] The current majority view is that invasion of South America by simians from Africa is at least as likely as entry from North America.[37,38] However, no platyrrhine fossil has yet been found in Africa. The early Oligocene parapithecids from the Fayum were previously suggested as platyrrhines[39] but are now considered to be early catarrhines.

The earliest accepted cercopithecoids (Old World monkeys) are *Prohylobates* and *Victoriapithecus*. Both genera are from early Miocene sites in Africa. North African *Prohylobates* is the most primitive and dates to about 18 Ma ago.[40-42] *Victoriapithecus* from several sites in East Africa is similar but slightly younger, and dates from a little over 16 to 15.5 Ma.[41,43] A molar and an isolated radio-ulna from Napak, Uganda may be 19 Ma old but this date is uncertain.[44] Neither *Prohylobates* nor *Victoriapithecus* can be ascribed to either of the modern Old World monkey families, Cercopithecidae and Colobidae.[19,45]

Another possible early cercopithecoid, *Dendropithecus*, dates to 17.5-20 Ma ago.[46] It was formerly described as a potential gibbon ancestor,[47] but its affinities are now uncertain.[32,45]

EARLY FOSSIL APES

Proconsul is an early Miocene catarrhine long known from several sites in East Africa. The genus shares some derived characters with all apes, but shows no particular affinity to any of the living apes. Hence, *Proconsul* has long been considered a potential ancestor for all living apes.[33,48,49] Harrison,[45] however, found little evidence to link *Proconsul* to apes and instead placed the genus as a sister group to all catarrhines. The most noted members of the genus are *P. africanus* and *P. nyanzae*, best known from Rusinga Island, Kenya where they date to 18 Ma. There are some possible *Proconsul* remains that date to 20-23 Ma. *Proconsul* may have appeared after the separation of apes and Old World monkeys, but before the separation of gibbons from great apes. If this interpretation is correct the oldest alleged proconsulids give the minimum date for the cercopithecoid/hominoid split of 20–23 Ma. The later 18 Ma date from Rusinga Island is consistent with the first appearance of the cercopithecoid *Prohylobates.*

The best evidence for the split between gibbons and great apes comes from fossils related to great apes as the gibbon fossil record is poor. There are only a few fragments of mid to late Miocene age that might belong to the gibbon lineage.[19] The earliest is a single unnamed upper molar from the Kamlial formation in the Siwaliks of northern Pakistan. It is dated to 16.1 Ma ago. *Dionysopithecus* from the Chinese mid to late Miocene (15-16 Ma) and the late Miocene *Krishnapithecus* from both India (7.4 Ma ago) and Mongolia may also be gibbons. These fragments are mostly thought to be gibbons because they occur in the right place at the right time. None of them actually shares any derived characters with gibbons.[50] *Pliopithecus* from mid-Miocene Europe (11-16 Ma), and its Chinese mid-Miocene relative *Laccopithecus,* were once thought to be gibbons,[51] but this view has been rejected. Pliopithecids are now thought to be either primitive apes or even a separate catarrhine branch arising before the ape-cercopithecoid split.[19,32]

Mid Miocene fossil apes from East Africa, notably *Afropithecus, Heliopithecus* and *Kenyapithecus*, share characteristics with great apes. This suggests that their appearance may set the minimum date for the gibbon/great ape split. *Afropithecus* is known from 16 to 17.2 Ma old sediments at Buluk, Kenya. *Heliopithecus,* from Saudi Arabia, is about the same age.[52-55] *Kenyapithecus* has aroused much interest as a putative common ancestor to living great apes and humans.[56,57] More than one hundred specimens are known from Kenya, ranging in age from 14 to 17 Ma.[57] The jaw of *Kenyapithecus* is ape-like though primitive and resembles the later European *Dryopithecus.*[33] However, *Kenyapithecus* is thought more similar to living great apes (i.e. less specialized) than *Dryopithecus,*[19,32,57,58] whose affinities within the great ape clade remain controversial.[59-63] Begun[59] suggests that it belongs with the African apes, while Solà and Köhler[63] assign it to the orangutan lineage. The earliest *Dryopithecus* dates to 14 Ma ago, about the same age as the latest *Kenyapithecus* from which it may have arisen. This is also about the same time as *Sivapithecus* appears in Asia.

Two decades ago most authorities regarded *Sivapithecus*—then called *Ramapithecus*—as ancestral to humans.[18,44,51] As more complete *Sivapithecus* specimens were discovered, and with the influence of molecular data discussed in chapter 5, this interpretation changed[64-67] to the current view that *Sivapithecus* is ancestral to orangutans only.[19,32,58,68-70] According to this model, the first appearance of *Sivapithecus* (12.5 Ma ago in the Siwalik Hills of Pakistan) is the latest possible fossil date for the division between *Pongo* and the African-ape (+ human) clade. However, an earlier date is indicated (14 Ma) if fragmentary remains from Turkey and Europe are accepted as *Sivapithecus*. This 14 Ma date would agree with the analysis of Begun[59] who places *Dryopithecus*, a contemporary of *Sivapithecus*, at the base of the African great apes. However, Kappelman et al[71] do not regard any of the Turkish or European fossils as belonging to *Sivapithecus*, as only scraps of crania have been recovered that lack derived traits unequivocally shared with orangutans.

The best known *Sivapithecus* species, *S. indicus*, does appear to share some derived cranial traits with orangutans[50,71] implying that they are sister groups. Pilbeam[56,72] earlier supported this view but now has doubts.[73,74] The similarity between *Sivapithecus* and the orangutan could be due to convergence. For instance, the large size and the masticatory and dental features shared by the two taxa may be due to a common diet and foraging style: the cracking of nuts, prolonged chewing, and ability to range widely and to use plants less commonly eaten by smaller primates.[75] Ward and Brown[70] support the placement of *Sivapithecus* in the *Pongo* lineage, but admit that convergent and parallel evolution in Miocene apes has been extensive and that other reconstructions are possible. Postcranial *Sivapithecus* remains analyzed by Pilbeam et al[74] show a number of primitive traits seen in cercopithecoids and *Kenyapithecus*, but which are not seen in living great apes (or humans). This result is reminiscent of earlier descriptions which likened *Sivapithecus* to some of the other Miocene African and European apes, and it calls into question yet again its affinities.[74] Kelley and Pilbeam[73] commented that: "... it can be argued that *Sivapithecus* is a sufficiently generalized ape ... to 'represent' comfortably the ancestors of all living hominoids".

If *Sivapithecus* and *Pongo* are sister taxa (the current orthodoxy) then some humeral features (e.g. markedly spool-shaped trochlea) shared by all great apes are convergent. On the other hand, if *Sivapithecus* is not part of the *Pongo* lineage but (like *Kenyapithecus*) is a sister group to all great apes, then the cranial similarities with *Pongo* are convergent. In such a case, the earliest appearance of *Sivapithecus* (12.5–?14 Ma) might antedate the *Pongo*/African-ape split.

Currently, the general view is that the known distribution and earliest occurrence of *Sivapithecus*—12.5 Ma in the Siwaliks, perhaps 14 Ma ago in Turkey—is consistent with an entry to Asia from Africa around 14–15 Ma ago (see chapter 9). In summary, *Sivapithecus* occurs at the right place and time to be a plausible, but not a definite, ancestor of *Pongo* with which it shares certain derived traits.

Ouranopithecus is a chimpanzee-sized fossil ape from the late Miocene of Greece and Macedonia.[76-77] It is broadly dated from 12 to 5.5 Ma ago. de Bonis et al[77] suggest that it may belong to the *Australopithecus-Homo* lineage, after separation from chimpanzees and gorillas, as it shares some derived dental traits with humans that are not seen in chimpanzees or gorillas. Andrews[78] questions the interpretation of de Bonis et al.[77] He points out that some of the presumed derived traits are seen in *Sivapithecus* and are therefore probably ancestral. Furthermore, if the examined fossil were female, rather than male as assumed, other traits are within the range of the African apes.

A maxilla from the Samburu Hills in Kenya appears to be similar to *Ouranopithecus*. This unnamed specimen has been dated at 9 Ma.[46] Its dentition is unlike any living ape, though its premolars are said to be gorilla-like.[48] Its dental enamel is very thick, like living humans but in contrast to thin enamel in living *Pan* and *Gorilla*, but thick enamel may be primitive.[79] Both *Ouranopithecus* and the Samburu maxilla may thus antedate the division between the human, chimpanzee and gorilla lineages and postdate the split between the orangutan and the African apes.

FOSSIL AFRICAN GREAT APES

There are no known Miocene or Pliocene fossils leading directly to *Gorilla* or *Pan*. The record of African Pliocene hominids related specifically to humans is much better, possibly because they preferred to live near lakes or rivers. The first characteristicly human skeletal characters to evolve were postcranial rather than cranial, judging by the earliest *Australopithecus*. They consist of leg and pelvic characters related to bipedalism and an upright posture. Therefore, the first appearance of hominoid bipedalism similar to that of humans is generally thought to indicate the latest date for the split between the human lineage and the other African apes.

The systematics of *Australopithecus* and *Homo* remain hotly debated. It is enough here to note that the earliest accepted members of *Australopithecus* had skeletal adaptations to a bipedal gait that somewhat resembled the gait of modern humans. It is generally regarded as unlikely that such specialized adaptations were found before the split, for that would imply that *Pan*, and perhaps *Gorilla* as well, had bipedal ancestors, and neither of these species shows any evidence of this.

However, opinions in paleoanthropology have changed radically in other matters. If *Pan* and/or *Gorilla* did actually have bipedal ancestors (a highly heterodox suggestion) then the earliest australopithecines antedate the split between humans and the African apes. This would allow much more recent human/African ape separation.[80]

There are a number of late Miocene and early Pliocene cranial fossils that may be assigned to *Australopithecus*. The earliest is a fragment of an adult right mandible retaining a worn molar from Lothagam,

Kenya, which dates to 5–5.5 Ma ago.[81] The affinities of the Lothagam mandible are uncertain. It appears to share some derived features with *Australopithecus afarensis*[82] but metrically it is between *Pan* and *Gorilla*. White[83] considers it to be primitive and does not classify it except as part of the African ape clade. Groves[19] considers it a hominin (that is, not on the *Pan* or *Gorilla* lineages), but Klein[24] suggests that it could have occurred either before the division of human and African apes, or afterwards but not on the human lineage.

In the Kanapoi formation, Kenya, a distal humerus ascribed to *A. afarensis* has been found, and dated to 4.01 ± 0.01 Ma ago. However, Feldersman[84] found on morphometric analysis that the Kanapoi humerus was most similar to that of a later hyper-robust australopithecine not considered to be in the direct human lineage.

The "Tabarin mandible" is a fragment of a small adult right dentary from the Chemeron formation at Tabarin, Lake Baringo, Kenya. It dates to between 4.96 ± 0.3 and 5.25 ± 0.04 Ma, and resembles the Lothagam find.[85,86] Ward and Hill[87] consider that in the limited parts preserved, the Tabarin mandible meets the definition of *A. afarensis*.[88]

These latest Miocene to early Pliocene fossil great apes from East Africa are fragmentary, so their ascription to *Australopithecus* is provisional. It is difficult to know their relationship to the human/African-ape separation in the absence of any postcranial remains.

EARLIEST FOSSIL HUMANS

Australopithecus is a known from many sites in East Africa and South Africa. The genus is paraphyletic because it is considered to include the ancestors of humans. Until recently the earliest fossils widely accepted as australopithecine were two Pliocene fragments, the Belohdelie frontal and the Maka proximal femur, from the Afar depression of Ethiopia, which both date to 3.9–4 Ma. White[89] ascribed both to *A. afarensis*. For the femoral fragment, bipedalism is probably indicated by the shape and the distribution of cortical thickening. However, Asfaw[90] remarked that the Belohdelie frontal is quite similar to that of extant African apes and that it might be "close to the divergence point for the hominid and African-ape clades". A well preserved *A. afarensis* skull (aged about 3 Ma) has recently been reported from Hadar, Ethiopia. This has a frontal very similar to the Belohdelie fragment, and supports the latter's australopithecine status.[91,92]

Many australopithecine fossils have been found at Laetoli, Kenya. These securely date to about 3.5–3.7 Ma ago,[19,93] because they are bracketed by tuffs dated radiometrically (Potassium-Argon) to 3.76 Ma below (Tuff 8) and 3.46 Ma above (Tuff 3). Far more dramatic are the bipedal footprints of three individuals (presumably *Australopithecus afarensis*) preserved in volcanic ash (Tuff 7) dated to 3.6–3.75 Ma ago.

A series of fossils ascribed to one or possibly two species of *Australopithecus* have been found in the Awash River valley of the Afar

Depression, Ethiopia. The Hadar site has provided the most primitive known hominins, which were bipedal yet lacked human-like dentition.[67] Feldersman[84] found their humeri to be intermediate in morphology between African apes and humans. The dates of the Hadar site have been disputed. Though initially said to be older than the Laetoli finds, the present consensus, from radiometric dates of associated tuffs, is that the Hadar site is slightly younger, aged about 3.3 Ma. Recent estimates range from 2.8 to 3.6 Ma.[94-98] A new site in southern Ethiopia, Fejej, near Lake Turkana, has recently produced some teeth that closely resemble *A. afarensis* teeth from Hadar and Laetoli. Their age, 3.6 Ma, is claimed to be at least as old as fossils from Laetoli.[93]

Stern and Susman[99] have described postcranial material from the Afar attributed to *A. afarensis* (some may belong in another taxon). They conclude that Hadar hominids AL 333 and AL 288 (the famous "Lucy") lived soon after the divergence and were still quite arboreal, though with a bipedal gait similar to, but not the same as, later *Homo*. Stern and Susman consider that *A. afarensis* from Hadar matches the footprints from Laetoli. The three individuals who made the Laetoli footprints had a bipedal gait quite similar to modern humans, though not exactly the same.[100]

There may have been more than one australopithecine species at Hadar. The Hadar skull AL 288-1 ("Lucy") in particular may represent *A. africanus*, not the more primitive *A. afarensis.*[101] *A. africanus*, first reported from South Africa, was a generalized mid-Pliocene bipedal hominid with a lifestyle unlike living humans or extant great apes. Lucy's bipedal gait may be derived and not intermediate between *Homo* and *Pan,* implying that the species including "Lucy" may not be ancestral to *Homo.*[102] Variation in Hadar humeri has also been alleged to indicate the presence of more than one species.[103]

Irrespective of the number of species at Hadar and of their precise relationship to later *Homo*, *Australopithecus* and modern *Homo* share enough derived traits associated with bipedalism to indicate separation of the human and African-ape lineages, assuming that bipedalism itself is derived. The separation must have occurred by 3.7 Ma based on the Laetoli footprints. The Tabarin mandible (4.96–5.25 Ma old) and most other fragments mentioned above cannot be assigned with certainty. If they are australopithecine an earlier separation is indicated. However, acceptance of the Belohdelie frontal and Maka femur (3.9–4 Ma old) as *Australopithecus* now seems reasonable.[91] This suggests a minimum date of about 4 Ma for the last common ancestor of chimpanzees and humans.

While this book was in preparation, a new fossil African ape, *Australopithecus ramidus*, was described from cranial and post-cranial remains found at Aramis in Ethiopia.[104,105] The maximum age is given as 4.4±0.03 Ma. The new species shows many chimpanzee-like features, and so is closer to the *Homo-Pan* split than is *A. afarensis.* White et al[104] consider that *A. ramidus* belongs on the *Homo* lineage after the split. This interpretation, which implies a new minimum age of about 4.4 Ma for the split, is supported by only three alleged australopithecine synapomorphies. These are the incisiform shape of the upper canine, a small protoconid on the otherwise chimpanzee-like deciduous lower molar,[105] and a relatively anterior position of the foramen magnum (suggesting upright posture). The possibility that *A. ramidus* antedates the divergence cannot be ruled out on the available evidence.

SUMMARY

We have presented a brief account of current opinion and interpretation of primate fossils as they relate to the divergence times of the taxa important in the context of the molecular clock debate. The results are summarized in Table 8.1. We repeat and emphasize our caveat from the first paragraph of this chapter: minimum divergence times derived from the fossil record may well change as new fossils are discovered and existing fossils are reinterpreted. The minimum divergence times estimated here will be discussed in chapter 10 in the context of deriving estimates of the rates of nucleotide substitution in both primates and other mammalian orders.

Table 8.1. Fossil evidence for minimun divergence times of pairs of primate taxa

		Taxon 1		Taxon 2	
Taxon 1	**Taxon 2**	**Earliest fossil**	**Earliest date (Ma)**	**Earliest fossil**	**Earliest date (Ma)**
Primates	other placentals	*Decoredon, ?Purgatorius*	60, ?67	not applicable	–
lorises+lemurs +galagos	tarsiers+monkeys +apes	*Cantius, Teilhardina*	55	*Altiatlasius*	57
tarsiers	monkeys+apes	*Nannopithex, Shoshonius*	50-55	*?Altiatlasius, Eosimias*	?57, 45
New World monkeys	Old World monkeys+apes	*Branisella*	26	*Amphipithecus, Pondaungia, Eosimias*	40 - 45
Old World monkeys	apes	*Prohylobates, Victoriapithecus*	18	*Proconsul*	18, ?20 - 23
gibbons	great apes	*?Dionysopithecus,*	16	*Kenyapithecus, Afropithecus, Heliopithecus*	17
orangutan	African apes (+human)	*Sivapithecus*	12.5 (14)	*Ouranopithecus,* Samburu maxilla	5.5 - 12,9
gorilla	chimpanzees +human	None known	–	*Australopithecus*	3.9 - 4
chimpanzees	human	None known	–	*Australopithecus*	?4.4

References

1. Simpson GG. The principles of classification and a classification of mammals. Bull Am Mus Nat Hist 1945; 85:1-350.
2. Novacek MJ, McKenna MC, Neff NA, Cifelli RL. Evidence from earliest known erinaceomorph basicranium that insectivorans and primates are not closely related. Nature 1983; 306:683-684.
3. Carroll RL. Vertebrate paleontology and evolution. New York: Freeman, 1988.
4. Martin RD. Primate origins and evolution, a phylogenetic reconstruction. London: Chapman & Hall, 1990.
5. Van Valen L, Sloan RE. The earliest primates. Science 1965; 150:743-745.
6. Romer AS. Vertebrate paleontology, 3rd edition. Chicago: University of Chicago Press, 1966.
7. Andrews P. A phylogenetic analysis of the primates. In: Benton MJ, ed. The phylogeny and classification of the tetrapods, volume 2: Mammals. Oxford: Clarendon Press, 1988:143-175.
8. Beard KC. Gliding behaviour and palaeoecology of the alleged primate family Paromomyidae (Mammalia, Dermoptera). Nature 1990; 345:340-341.
9. Kay RF, Thorington RW, Houde P. Eocene plesiadapiform shows affinities with flying lemurs, not primates. Nature 1990; 345:342-344.
10. Krause DW. Were paromomyids gliders? Maybe, maybe not. Hum Evol 1991; 21:177-188.
11. Martin RD. Primate origins and evolution, a phylogenetic reconstruction. London: Chapman and Hall, 1990.
12. Martin RD. Some relatives take a dive. Nature 1990; 345:291-292.
13. Szalay FS, Li C-K. Middle Palaeocene euprimate from southern China and the distribution of Primates in the Palaeogene. J Hum Evol 1986; 15:387-397.
14. Sige B, Jaeger J, Sudre J, Vianey-Liaud M. *Altiatlasius kolchii*, n., gen et sp., primate omonyidé du Paléocène supérieur du Maroc, et les origines des euprimates. Palaeontographica Abt A 1990; 214:31-56.
15. Beard KC, Qi T, Dawsin MR, Wang B, Li C. A diverse new primate fauna from middle Eocene fissure-fillings in southeastern China. Nature 1994; 368:604-609.
16. Szalay FS, Delson E. Evolutionary history of the primates. New York: Academic Press, 1979.
17. Rose KD, Walker A. The skeleton of early Eocene *Cantius*, oldest lemuriform primate. Am J Phys Anthrop 1994; 66:73-89.
18. Simons EL. The fossil record of primate phylogeny. In: Goodman M, Tashian RE, eds. Molecular anthropology. New York: Plenum Press, 1976:35-62.
19. Groves CP. A theory of human and primate evolution. Oxford: Oxford University Press, 1989.
20. Godinot M, Russell DE, Louis P. Oldest known *Nannopithex* (Primates,

Omomyiformes) from the early Eocene of France. Folia Primatol 1992; 58:32-40.
21. Beard KC, Krishtalka L, Stucky RK. First skulls of the early Eocene primate *Shoshonius cooperi* and the anthropoid-tarsier dichotomy. Nature 1991; 349:64-67.
22. Maw B, Ciochon RL, Savage DE. Late Eocene of Burma yields earliest anthropoid primate. Nature 1979; 282:65-67.
23. Ciochon RL, Savage DE, Tint T, Maw B. Anthropoid origins in Asia? New discovery of *Amphipithecus* from the Eocene of Burma. Science 1985; 229:756-759.
24. Klein RG. The human career: human biological and cultural origins. Chicago: University of Chicago Press, 1989.
25. Simons EL. Relationships of *Amphipithecus* and *Oligopithecus*. Nature 1971; 232:489-491.
26. Simons EL. Description of two genera and species of late Eocene Anthropoidea from Egypt. Proc Natl Acad Sci (USA) 1989; 86:9956-9960.
27. Van Couvering JA, Harris JA. Late Eocene age of Fayum mammal faunas. J Hum Evol 1991; 21:241-260.
28. Gingerich PD. Oligocene age of the Gebel Qatrani Formation, Fayum, Egypt. J Hum Evol 1993; 24:207-218.
29. Simons EL. The earliest apes. Sci Am 1967; 217:28-35.
30. Simons EL. Natural History. Nat Hist 1984; 93:18-20.
31. Simons EL. Origins and characteristics of the first hominoids. In: Delson E, ed. Ancestors: the hard evidence. New York: Alan Liss, 1985:37-41.
32. Andrews P. Family group systematics and evolution among catarrhine primates. In: Delson E, ed. Ancestors: the hard evidence. New York: Alan Liss, 1985:14-20.
33. Delson E. Catarrhine evolution. In: Delson E, ed. Ancestors: the hard evidence. New York: Alan R. Liss, 1985:9-13.
34. Butler PM. Problems of dental evolution in the higher primates. In: Wood B, Martin L, Andrews P, eds. Major topics in primate and human evolution. Cambridge: Cambridge University Press, 1986:89-106.
35. Fleagle JC. The fossil record of early catarrhine evolution. In: Wood B, Martin L, Andrews P, eds. Major topics in primate and human evolution. Cambridge: Cambridge University Press, 1986:130-149.
36. Hershkovitz P. A new genus of late Oligocene monkey (Cebidae, Platyrrhini) with notes on postorbital closure and platyrrhine evolution. Folia Primatol 1974; 21:1-35.
37. Ciochon RL, Chiarelli AB. Evolutionary biology of the New World monkeys and continental drift. New York: Plenum Press, 1980.
38. MacFadden BJ. Chronology of Cenozoic primate localities in South America. J Hum Evol 1990; 19:7-21.
39. Hoffstetter R. Origins and deployment of New World monkeys emphasizing the southern continents route. In: Ciochon RL, Chiarelli AB, eds. Evolutionary biology of the New World monkeys and continental drift. New York: Plenum Press, 1980:103-138.

40. Simons EL. Miocene monkey (*Prohylobates*) from northern Egypt. Nature 1969; 223:687-689.
41. Delson E. *Prohylobates* (Primates) from the early Miocene of Libya: a new species and its implications for cercopithecoid origins. Geobios 1979; 12:725-733.
42. Fleagle JG. Early anthropoid evolution in Africa and South America. In: Else JG, Else PC, eds. Primate evolution. Cambridge: Cambridge University Press, 1986:133-142.
43. Pickford M. The chronology of the Cercopithecoidea of East Africa. Hum Evol 1987; 2:1-17.
44. Pilbeam D, Walker A. Fossil monkeys from the Miocene of Napek, North-East Uganda. Nature 1968; 220:657-660.
45. Harrison T. The phylogenetic relationships of the early catarrhine primates: a review of the current evidence. J Hum Evol 1987; 16:41-80.
46. Pickford M. The geochronology of Miocene higher faunas of East Africa. In: Else JG, Else PC, eds. Primate evolution. Cambridge: Cambridge University Press, 1986:19-33.
47. Andrews P, Simons E. A new African Miocene gibbon-like genus, *Dendropithecus* (Hominoidea, Primates) with distinctive postcranial adaptations: its significance to origin of Hylobatidae. Folia Primatol 1977; 28:161-169.
48. Pilbeam D. Hominoid evolution and hominoid origins. Am Anthrop 1986; 88(3):295-312.
49. Andrews P, Martin L. Cladistic relationships of extant and fossil hominoids. J Hum Evol 1987; 16:101-118.
50. Barry JC. A review of the chronology of Sivalik hominoids. In: Else JG, Else PC, eds. Primate evolution. Cambridge: Cambridge University Press, 1986:93-106.
51. Walker A. Splitting times among hominoids deduced from the fossil record. In: Goodman M, Tashian RE, eds. Molecular anthropology. New York: Plenum Press, 1976:63-77.
52. Leakey REF, Walker A. New higher primates from the early Miocene of Buluk, Kenya. Nature 1985; 318:173-175.
53. Delson E. The earliest *Sivapithecus*? Nature 1985; 318:107-108.
54. McDougall I, Watkins RT. Age of hominoid-bearing sequence at Buluk, northern Kenya. Nature 1985; 318:175-178.
55. Leakey REF, Leakey MG, Walker AC. Morphology of *Turkanopithecus kalakolensis* from Kenya. Am J Phys Anthrop 1988; 76:277-288.
56. Pilbeam D. Patterns of hominoid evolution. In: Delson E, ed. Ancestors: the hard evidence. New York: Alan Liss, 1985:51-59.
57. Pickford M. A reappraisal of *Kenyapithecus*. In: Else JG, Else PC, eds. Primate evolution. Cambridge: Cambridge University Press, 1986:163-171.
58. Martin L. Relationships among extant and extinct great apes and humans. In: Wood B, Martin L, Andrews P, eds. Major topics in primate and human evolution. Cambridge: Cambridge University Press, 1986:151-187.
59. Begun DR. Miocene fossil hominids and the chimp-human clade. Sci-

ence 1992; 257:1929-1932.
60. Gibbons A. Hungarian fossils stir debate on ape and human origins. Science 1992; 257:1864-1865.
61. Dean D, Delson E. Second gorilla or third chimp? Nature 1992; 359:676-677.
62. Martin L, Andrews P. Renaissance of Europe's ape. Nature 1993; 365:494.
63. Solá SM, Köhler M. Recent discoveries of *Dryopithecus* shed new light on evolution of great apes. Nature 1993; 365:543-545.
64. Preuss TM. The face of *Sivapithecus indicus*; description of a new relatively complete specimen from the Siwaliks of Pakistan. Folia Primatol 1982; 38:141-157.
65. Raza SM, Barry JC, Pilbeam D, Rose MD, Shah SMI, et al. New hominoid primates from the middle Miocene Chinji formation, Potwar Plateau, Pakistan. Nature 1983; 306:52-54.
66. Sarich VM. Retrospective on hominoid macromolecular systematics. In: Ciochon RL, Corruccini RS, eds. New interpretations of ape and human ancestry. New York: Plenum, 1983:137-150.
67. Pilbeam D. The descent of hominoids and hominids. Sci Am 1984; 250:84-96.
68. Andrews P. Hominoid evolution. Nature 1982; 295:185-186.
69. Andrews P, Cronin JE. The relationships of *Sivapithecus* and *Ramapithecus* and the evolution of the orang-utan. Nature 1982; 297:541-545.
70. Ward SC, Brown B. The facial skeleton of *Sivapithecus indicus*. In: Swindler DR, Erwin J, eds. Comparative primate biology. New York: Alan Liss, 1986:413-452.
71. Kappelman J, Kelley J, Pilbeam D, Sheikh KA, Ward S, et al. The earliest occurence of Sivapithecus from the middle Miocene Chinji formation of Pakistan. J Hum Evol 1991; 21:61-73.
72. Pilbeam D. New hominoid skull material from the Miocene of Pakistan. Nature 1982; 295:232-234.
73. Kelley JK, Pilbeam D. The dryopithecines: taxonomy, comparative anatomy, and phylogeny of Miocene large hominoids. In: Swindler DR, Erwin J, eds. Comparative primate biology, Vol. 1. New York: Alan Liss, 1986:361-411.
74. Pilbeam D, Rose MD, Barry JC, Ibrahim Shah SM. New *Sivapithecus* humeri from Pakistan and the relationship of *Sivapithecus* and *Pongo*. Nature 1990; 348:237-239.
75. Wheatley BP. The evolution of a large body size in orangutans: a model for hominoid divergence. Am J Primatol 1987; 13:313-324.
76. de Bonis L, Bouvrain G, Koufos G, Melentis J. Succession and dating of the late Miocene primates of Macedonia. In: Else JG, Else PC, eds. Primate evolution. Cambridge: Cambridge University Press, 1986:107-114.
77. de Bonis L, Bouvrain G, Geraads D, Koufos G. New hominid skull material from the late Miocene of Macedonia in northern Greece. Nature 1990; 345:712-714.
78. Andrews P, Alpagut B. Description of the fossiliferous units at Pasalar,

Turkey. J Hum Evol 1990; 19:343-361.
79. Martin L. Significance of enamel thickness in hominoid evolution. Nature 1985; 314:260-263.
80. Verhaegen MJB. African ape ancestry. Hum Evol 1990; 19:295-297.
81. Patterson B, Behrensmeyer AK, Sill WD. Geology and fauna of a new Pliocene locality in north-western Kenya. Nature 1970; 226:918-921.
82. Kramer A. Hominid-pongid distinctiveness in the Miocene-Pliocence fossil record: the Lothagam mandible. Am J Phys Anthrop 1986; 70:457-473.
83. White TD. *Australopithecus afarensis* and the Lothagam mandible. Anthropos 1986; 23:79-90.
84. Feldersman MC. Morphometric analysis of the distal humerus of some Cenozoic catarrhines: the late divergence hypothesis revisited. Am J Phys Anthrop 1982; 59:73-95.
85. Hill A. Early hominid from Baringo, Kenya. Nature 1985; 315:222-224.
86. Hill A, Drake R, Tauxe L, Monaghan M, Barry JC, et al. Neogene palaeontology and geochronology of the Baringo Basin, Kenya. J Hum Evol 1985; 14:759-773.
87. Ward S, Hill A. Pliocene hominid partial mandible from Tabarin, Baringo, Kenya. Am J Phys Anthrop 1987; 72:21-37.
88. Johanson DC, Shreeve J. Lucy's child. The discovery of a human ancestor. London: Penguin, 1989.
89. White TD. Pliocene hominids from the Middle Awash, Ethiopia. Courier Forschungsinstitut Senckenberg 1984; 69:57-68.
90. Asfaw B. The Belodelie frontal: a new evidence of early hominid cranial morphology from the Afar of Ethiopia. J Hum Evol 1987; 16:611-624.
91. Kimbel WH, Johanson DC, Rak Y. The first skull and other new discoveries of *Australopithecus afarensis* at Hadar, Ethiopia. Nature 1994; 368:449-451.
92. Shreeve J. 'Lucy', crucial early human ancestor, finally gets a head. Science 1994; 264:34-35.
93. Wood B. A remote sense for fossils. Nature 1992; 355:397-398.
94. Walter RC, Aronson JL. Revisions of K/Ar ages for the Hadar hominid site, Ethiopia. Nature 1982; 296:122-123.
95. Brown FH. Tulu Bor Tuff at Koobi fora correlated with the Sidi Hakoma Tuff at Hadar. Nature 1982; 300:631-633.
96. Boaz NT, Howell FC, McCrossin ML. Faunal age of the Usno, Shungura B and Hadar formations, Ethiopia. Nature 1982; 300:633-635.
97. Schmitt TJ, Nairn EM. Interpretations of the magnetostratigraphy of the Hadar hominid site, Ethiopia. Nature 1984; 309:704-706.
98. Sarna-Wojcicki AM, Meyer CE, Roth PH, Brown FH. Ages of tuff beds at East African early hominid sites and sediments in the Gulf of Aden. Nature 1985; 313:306-308.
99. Stern JT, Susman RL. The locomotor anatomy of *Australopithecus afarensis.* Am J Phys Anthrop 1983; 60:279-317.
100. Day MH, Wickens EH. Laetoli Pliocence hominid footprints and bipedalism. Nature 1980; 286:385-387.

101. Schmid P. A reconstruction of the skeleton of A.L. 288-1 (Hadar) and its consequences. Folia Primatol 1983; 40:283-306.
102. Rak Y. Lucy's pelvic anatomy: its role in bipedal gait. J Hum Evol 1991; 20:283-290.
103. Senut B, Tardieu C. Functional aspects of Plio-Pleistocene hominid limb bones: implications for taxonomy and phylogeny. In: Delson E, ed. Ancestors: the hard evidence. New York: Alan Liss, 1985:193-201.
104. White TD, Suwa G, Asfaw B. Australopithecus ramidus, a new species of early hominid from Aramis, Ethiopia. Nature 1994; 371:306-312.
105. Wood B. The oldest hominid yet. Nature 1994; 371:280-281.

CHAPTER 9

BIOGEOGRAPHY AND THE DIVERGENCE OF PLACENTAL MAMMALS

Treatments of placental phylogeny, whether based on anatomy or molecules, often pay little attention to biogeography. Norell and Novacek[1] for instance mention geographic factors, but go on to state that fossils are "the only direct source of evidence of past organismic history". Fossils are indeed direct evidence for the *existence* of past organisms, but this does not mean that they provide the only direct evidence of the *evolutionary history* of those organisms. Evidence of evolutionary history, be it from fossils, morphology, molecules or geography, can only be obtained through a veil of interpretation. It is misleading to suggest that fossils have the dominant position in the hierarchy of such evidence, or that fossils should serve as the sole benchmark for the evaluation of all other forms of evidence.

A more useful approach is to consider all forms of available evidence, with their limitations, and thereby attempt to derive the best estimate currently possible. We showed in chapter 7 that divergence of placental orders well before the Cretaceous–Tertiary boundary was compatible with, but not necessarily predicted by, the incomplete fossil record. Paleobiogeography provides a third form of evidence about divergence times and thus may be useful in resolving incompatibilities between paleontological and molecular interpretations. Since conclusions from molecular data generally depend on the assumption of a molecular clock, paleobiogeography has an important bearing on whether the molecular clock hypothesis should be rejected. Appreciation of its role in this regard requires an understanding of the Mesozoic and Cenozoic pattern of continental drift.

The following brief account is derived mainly from geophysicists[2-9] although we have also taken note of the comments of biologists.[10-13] The positions of continental land masses in the late Jurassic, in the late Cretaceous and in the Oligocene are shown in Figures 9.1, 9.2 and 9.3 respectively.

CONTINENTAL DRIFT

The gross geography of the earth appears to follow a 400-500 Ma tectonic and climatic cycle. The extremes are a single supercontinent (Pangea) with a single world-ocean (Panthalassa) under an "icehouse" climate, and many dispersed continents with a "greenhouse" climate (as at present). About 180 Ma ago (mid-Jurassic), the latest Pangea began to break up, with the Tethys Sea separating a northern supercontinent (Laurasia) from a southern supercontinent (Gondwana or Gondwanaland). This division was completed by 140 to 118 Ma ago.[8] The Tethys was then at its widest and was fully open westwards to the proto-Atlantic. Africa and the rest of Gondwana were completely separate from Eurasia and North America, together known as Laurasia.

About 85 Ma ago Africa (formerly in western Gondwana) began to move northward relative to Europe, gradually closing the western Tethys in the vicinity of the present day western Mediterranean. Closure dates are debated; complete closure may have been late Oligocene or early Miocene, although there may have been earlier intermittent land bridges from North Africa to Europe.[9,14] From the mid-Cretaceous to the Paleocene, West Africa was isolated from the rest of Africa by an epicontinental Trans-Saharan Seaway. This was associated with a series of rift valleys radiating from the opening of the Atlantic Ocean between Africa and South America.[12,15,16]

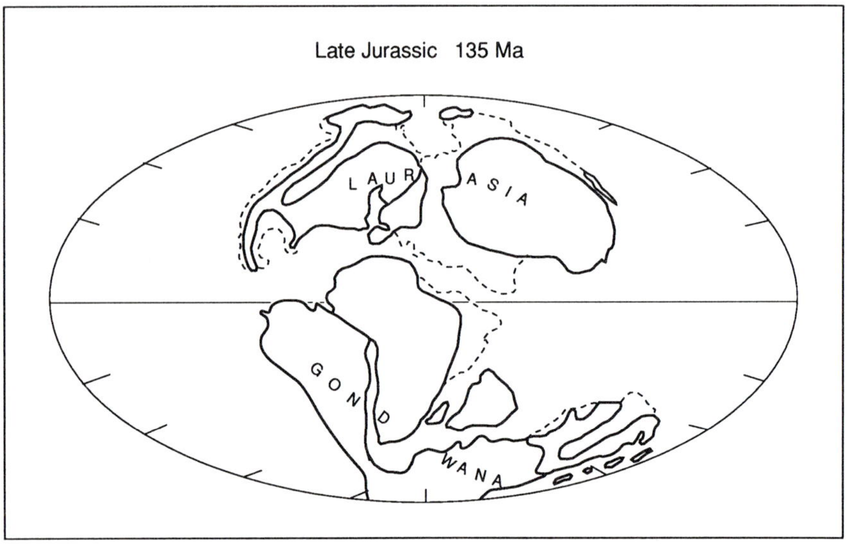

Fig. 9.1. Land mass distributions in the late Jurassic. Modified and reprinted with permission from LA Frakes and PA Vickers-Rich. In: Vickers-Rich et al, eds. Vertebrate paleaontology of Australasia. Melbourne: Monash University Publications, 1991:111-145.

To the east, land bridges developed between Africa and Asia across the narrowing Tethys. Mammals apparently entered and left Africa after the late Oligocene. Land bridges between Africa and Eurasia became briefly established in the early to mid-Miocene, but did not become permanent until much later. One arm of the narrowing Tethys ran through the present Persian Gulf and Iraq into the proto-Mediterranean for most of the early to mid-Miocene, blocking land routes between Africa and Eurasia. During the mid-Miocene, this gap was closed. Extensive faunal exchange appears to have occurred in two distinct "Neogene Dispersal Phases" (NDPs). These have been dated to 18 ± 1 Ma (NDP 1) and to 15 ± 1 Ma (NDP 2),[17,18] and are discussed below in relation to primate evolution.

North and South America were separated by ocean by at least 140 Ma ago and possibly earlier.[11] The Americas remained separate until the Pliocene emergence of the Isthmus of Panama, about 3 Ma ago. Islands may have been present before this time, providing some potential for faunal exchange.

South America lay more to the east of North America than presently. As the prevailing equatorial current was westwards, dispersal from South America to North America would have been more favored than crossing in the reverse direction.[3] This may contradict some biogeographic models that attempt to relate South American therian taxa to known Paleogene North American fossils by direct dispersal southwards.[19]

North America and Europe were formerly joined as so-called Euramerica, to the north of a developing North Atlantic. The Atlantic began to open in Jurassic times, about 180 Ma ago. The timing of the final severing is disputed. Some plate reconstructions suggest a seaway as early as the mid-Mesozoic, but Irving[2] and Weijermars[9] show it as still closed at the end of the Cretaceous, and paleontological evidence suggests that the North Atlantic land bridge did not close until the early Eocene.[20]

The present Eurasia did not exist in the Mesozoic. There was no South Asian extension until the early Eocene, when the formerly isolated Indian plate (a fragment of Gondwana) collided with Eurasia. There may have a dispersal route between them available earlier, but no Holarctic fauna is known in India until the end of the Cretaceous.[21]

Eurasia was also divided in the later Mesozoic by a north-south epicontinental seaway, the Turgai Sea (also called the Turgai Straits and the Obik Sea). This lay east of the present-day Urals, and divided Asia and Europe for 130 Ma, from the late Jurassic till the Oligocene. Latterly, the barrier may have been intermittent or incomplete. The Turgai Sea linked the proto-Arctic Ocean to a northern branch of the Tethys, the remnants of which persist as the Black, Caspian and Aral Seas. The major Oligocene extinction in Western Europe known as the "*Grand Coupure*" coincides with the end of the Turgai Sea and the reunification of Eurasia.

Thus by the mid-Cretaceous there were two northern continents, Asia to the east of the Turgai Sea and Euramerica to the west. Soon after, during the middle to late Cretaceous, North America was itself divided by the north-south Skull Creek or Mid-Continental Seaway. There were then effectively three northern continents: Asia, Western North America, and Euramerica. This suggests why the earliest Tertiary fossil fauna of Eastern Eurasia differs from that known in Europe and North America.

In the late Cretaceous, the site of the present Bering Straits closed. This produced a land bridge, "Beringia", between Asia and Western North America that has been intermittently present through much of the Cenozoic. By the end of the Cretaceous, the north-south Skull Creek Seaway of North America had dried up: this effectively produced a single northern continent, which until the Oligocene was interrupted only by the Turgai Sea. In the late Paleocene, North American fauna came to resemble that of Eastern Asia, confirming the presence of Beringia at least intermittently. This model is consistent with recognition by biogeographers (since the time of Wallace) of a large scale Holarctic region.

The ancient southern supercontinent, Gondwana, antedates the latest Pangea. Gondwana persisted largely intact from the latest Precambrian breakup of an earlier Proto-Pangea to the reformation of Pangea 320 Ma ago. Until 160 Ma ago, all the present southern continents were thus connected. From 160 Ma ago, western Gondwana (Africa, Madagascar, South America) began to separate from eastern Gondwana (the Indian subcontinent, Antarctica, Australia, New Zealand, New Caledonia). The rift between India and Australia-Antarctica appeared by 128 Ma ago and was fully open by 118 Ma ago. At much the same time, South America began to rift from Africa.

From about 130 Ma to 119 Ma ago, the South Atlantic Ocean was opening from the south northwards, separating South America from Africa. From about 119 Ma to 105 Ma ago South America moved westward relative to Africa: the future Equatorial Atlantic became a complex series of basins. The deep ocean basins finally merged about 85 Ma ago.[15,22] It seems unlikely that a complete land bridge remained between West Africa and northeastern South America after about 100-105 Ma ago. Volcanic islands may have provided a ready dispersal route for another 20 Ma.

By 80-85 Ma ago, Africa and South America were nowhere less than 500 km apart. To the south, large mid-ocean islands probably existed, particularly along the Walvis-Rio Grande Rise: such islands were the tectonic equivalent of modern Iceland, and were surrounded by deep ocean. To the north, large volcanic islands probably lay between what are now Sierra Leone and Brazil, but by the late Cretaceous even the narrowest gap in such a chain was probably many hundreds of kilometers wide. Dispersal from Africa to South America was, however, favored by equatorial ocean currents from the east.

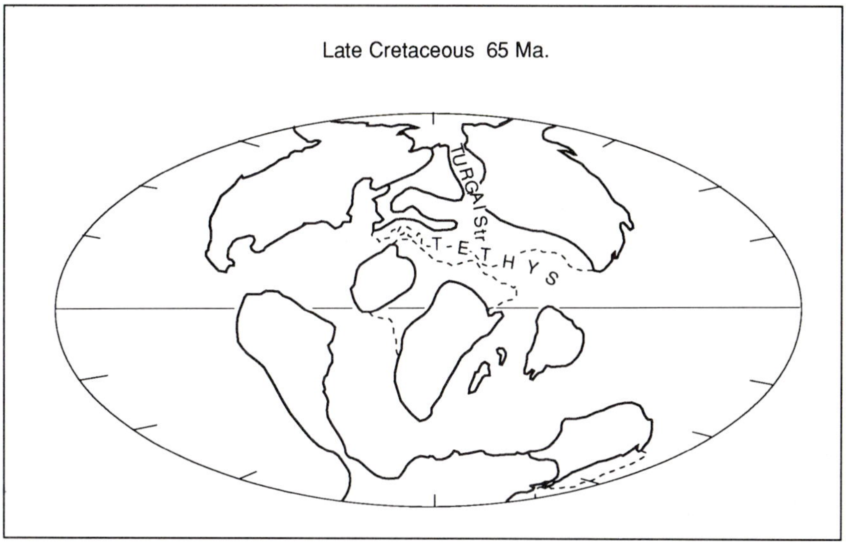

Fig. 9.2. Land mass distributions in the late Cretaceous. Modified and reprinted with permission from LA Frakes and PA Vickers-Rich. In: Vickers-Rich et al, eds. Vertebrate paleaontology of Australasia. Melbourne: Monash University Publications, 1991:111-145.

The relationships between Africa, Madagascar and India are complex. Most reconstructions suggest that, relative to Africa, Madagascar was formerly more to the north and adjacent to present Somalia. Rifting southwards commenced about 165 Ma ago and ceased 121 Ma ago. Since the early Cretaceous, Madagascar has thus been at about the same position relative to Africa.[23] The ocean between India and Madagascar, although wider, is more recent. While Madagascar moved south, the Indian subcontinent moved north. The continental shelves, at least, of India and Madagascar were still adjoining 140 Ma ago, and possibly later (with a possible Indian link to Africa as late as the end of the Cretaceous), but there was probably full separation by 95 Ma ago.

The fate of eastern Gondwana is equally complex. In Southeast Asia, numerous terrains mainly derive from Gondwana. However, rifting of most of these had occurred by the late Triassic[24,25]—probably too early to be of relevance to vicariant distribution of therians. During the later Mesozoic, Australia moved north to eventually collide with the Philippines and China plates in the Miocene.

Antarctica has remained relatively constant in position since the Mesozoic. Australia and Antarctica began to separate in the mid-Jurassic, some 160 Ma ago. This was initially by continental extension, but from about 96 Ma ago seafloor began to develop between the continents. There was apparently a complete gap by 35-45 Ma ago. A spreading ridge between southeastern Australia and New Zealand developed about 95 Ma ago, with deep ocean present by 82 Ma ago. The date of final separation of New Zealand from Antarctica is less clear, but may also

Fig. 9.3. Land mass distributions in the Oligocene. Modified and reprinted with permission from LA Frakes and PA Vickers-Rich. In: Vickers-Rich et al, eds. Vertebrate paleaontology of Australasia. Melbourne: Monash University Publications, 1991:111-145.

have been this early. This model of an ancient separation is consistent with the great difference between the Australian and New Zealand biotas.

The separation of South America from Antarctica was much later than that of the former from North America or Africa. The oldest ocean crust in Drake's passage seems at the oldest to be either latest Oligocene or early Miocene, suggesting a definite gap by 20 Ma ago at the latest.[2,6,9] This complete isolation of Antarctica marked the final breakup of Gondwana.

DISTRIBUTION OF LAND MASSES AT THE END OF THE MESOZOIC

Sixty-five million years ago, world geography was clearly different than it is now. At the time of the Cretaceous-Tertiary mass extinction, Laurasia probably formed a unit. East Asia was separated by water from Western Eurasia (by the Turgai Sea), but probably not from North America (the Beringia land bridge). North America was also probably still connected by land to Europe.

To the south, Gondwana had been separated from Laurasia for more than 50 million years and the Americas had been divided by sea since 140 Ma ago. South America, Antarctica and Australia were still a unit, but Africa, the Indian subcontinent, Madagascar, New Caledonia and New Zealand were all well isolated by ocean. Africa had long been isolated from Laurasia (118 Ma ago), South America (105 Ma ago), Madagascar (perhaps 160 Ma ago) and India (perhaps 95 Ma ago). Africa may have had a northwest land bridge to Europe in the earliest Paleocene and again in the late Paleocene. However, the west of Africa was at that time isolated from the rest of the continent by an epicontinental sea (Trans-Saharan Seaway).

BIOGEOGRAPHY AND THERIAN PHYLOGENY

It seems likely that marsupials[12,27-29] and possibly monotremes[28,30] diverged before the Paleocene or late Cretaceous, based on their present distribution. A similar argument can be made for some placental mammals including "edentates", rodents and primates.

The Xenarthra (armadillos, anteaters and sloths) are now found only in South America, except for armadillos which invaded North America in the Pleistocene. The Pholidota (pangolins) are confined to southern Asia and Africa. Yet both xenarthrans and pholidotes lived in Eocene Europe.[31,32] By the Eocene, the Americas, Europe, Africa and South Asia had no land bridges between them. A plausible model for the xenarthran distribution involves vicariance, with xenarthrans already living in Europe, South America and Africa at a time when these continents were still connected by dry land. This requires an origin of the order more than 100 Ma ago.[12]

The alternative explanation for the presence of these taxa in Eocene Europe seems highly improbable. It involves a northwards crossing by sea from South America to North America, then entry to Europe via one of the two land bridges to Eurasia. This implies a xenarthran presence in Paleogene North America for which there is no evidence.

Similar considerations apply to the Pholidota, the earliest known representative of which is the Eocene *Eomanis* from Europe.[31] There was apparent no dry land route between Europe, South Asia and Africa until the Oligocene, as these regions were separated by the Turgai Sea and Tethys Ocean. A vicariant explanation of the occurrence of pholidotes in these regions implies an origin of the order no later than mid Cretaceous.

A large number of mammalian orders peculiar to South America flourished there throughout most of the Tertiary.[20,33] Xenarthrans and the three extant neotropical marsupial clades are merely the surviving examples. This pattern is consistent with prolonged isolation of South America from North America (perhaps 140 Ma) and from Africa (perhaps 105 Ma). However, De Muizon and Marshall[34] report a primitive pantodont (early placental) from the early Paleocene of Bolivia. Previously, this group was known only from the northern continents. The authors would prefer a dispersal southwards. An alternate explanation is that pantodonts were established by the early Cretaceous.

Rodentia include more species than all other therian orders combined, and have a matching complexity in taxonomy. Rodents are now commonly divided into two sub-orders: Hystricognatha (porcupines, guinea pigs and their relatives) and Sciurognatha (mice, squirrels etc.). Hystricognaths present two biogeographic problems if rodents evolved as late as the time of the mass extinction. First, their oldest fossils are

Eocene and in Africa, which at that time was isolated from Europe (and everywhere else). If hystricognaths had evolved in the Paleocene from nonhystricognaths in Asia (the putative origin of all rodents), their only feasible land route to Eocene Africa was eastwards via Beringia to North America, thence to Europe and thence to North Africa during one of the Paleocene dispersal phases. Yet Paleocene hystricognaths are quite unknown from either North America or Europe. Direct spread westwards from Asia to Africa was blocked by the Turgai Sea and Tethys Ocean, until the Oligocene and Miocene respectively.

Second, hystricognaths are divided into two geographic groups by the Atlantic Ocean: Caviomorpha (guinea pigs etc.) in the New World (mainly South America) and Phiomorpha (Old World porcupines and relatives), mainly in Africa. This split has long been recognized as parallel to the trans-Atlantic separation of simian primates into platyrrhines (New World monkeys) and haplorrhines (Old World monkeys).[35] If the hystricognath (and simian) separation arose strictly by vicariance, the latest likely date is 85–105 Ma ago. Later dates require a difficult ocean crossing from Africa to South America. The entry of caviomorphs to South America from North America seems unlikely for geographic reasons stated earlier. Recently, a primitive caviomorph has been discovered in late-Eocene to early-Oligocene deposits in Chile. This find has increased the known age of caviomorphs by 10 Ma.[36]

Biogeographic considerations thus suggest that the separation of hystricognath from sciurognath rodents, and possibly of caviomorphs from phiomorphs, had already occurred by the mid-Cretaceous. If this was the case, the separation of rodents from other placental orders must have occurred at least 100 Ma ago, possibly much earlier.

There is a recent suggestion, based on molecular data, that rodents are polyphyletic,[37-39] although more recent analysis of molecular data refutes this.[40-42] If it were true, hystricognaths would belong to a new order separate from other rodents, and comprise an outgroup to a placental clade including at least primates, artiodactyls and other rodents. In either case, biogeography implies a pre-mid Cretaceous divergence of hystricognaths from other rodents.

Judging by past and present distributions, African origins seem likely for hyracoids (hyraxes), macroscelideans (elephant shrews), tubulidents (aardvark), proboscideans (elephants), and primates. The first three orders are unknown elsewhere before the Miocene. By contrast, Laurasian origins seem likely for carnivores, dermopterans (flying lemurs), lagomorphs (rabbits and hares), lipotyphlans (insectivores) and scandentians (tree shrews). At the time of the Cretaceous–Tertiary mass extinction, Africa and Laurasia had been separated for 50 Ma. This would suggest that these two groups of placental orders separated from each other more than 100 Ma ago.

Primate biogeography is discussed in some detail below, where we note that there is suggestion of a distinct primate lineage by 85-100 Ma ago. In the present context, we also point out that the origin of the present distribution of strepsirhines (galagos, lorises and lemurs) would be greatly simplified by assuming a mid-Cretaceous vicariant split from the haplorhines (tarsiers, monkeys and apes). Likewise, the distribution of tarsiers may be explained by descent from omomyiforms entering Euramerica from Africa in the Paleocene or perhaps as early as the mid-Cretaceous. So divergences may have occurred even within primate sub-orders before the end of the Cretaceous.

These biogeographical considerations indicate that extant placental orders do not represent a mass radiation from one or two lineages surviving the Cretaceous-Tertiary mass extinction. Instead, living placentals represent many clades that independently survived the mass extinction in scattered parts of the world. A star-like mass radiation of placental orders some 65 Ma ago has been a convenient assumption, but one based largely on a relative paucity of Cretaceous fossil evidence. It seems much more likely that the pattern of branching in the lineages leading to the living orders of placentals and marsupials was hierarchical and dichotomous, and that divergences began at least as early as the mid-Cretaceous or even late Jurassic, 100 to 150 Ma ago.

PRIMATE BIOGEOGRAPHY

Geographical Origins of Primates

In a review of new fossil evidence, Martin[43] repeats his earlier challenges[12] to traditional views of primate origins. The order was previously thought to have originated during the late Cretaceous or Paleocene in either Europe or North America (which were then linked). This was based on the view that modern primates evolved from the Laurasian plesiadapiforms. North Africa has now revealed a series of Paleocene or early Eocene "euprimates" (primates that share modern features). These finds imply more ancient separations within the order, and, by implication, even earlier separations of primates from other placentals. In addition, the past and present distributions of primates make Africa a far more plausible center for their evolution than Asia, Europe or North America.[43-46]

Haplorhines

Simians probably evolved somewhere in Gondwana (probably Africa where their past and present diversity is greatest) and later spread to Laurasia. The only reliable early Tertiary fossils of simians from the northern continents are *Pondaungia* and *Amphipithecus* from Eocene Burma and *Eosimias* from Eocene China. A Laurasian origin of primates would require spread from North America and/or Europe or Asia to both Africa and to South America. Living primates are mainly tropical and subtropical, making high latitude dispersal routes less likely. Due to lack of land bridges dispersal out of Laurasia into Africa or

North America would have been geographically very difficult. From the Eocene until late Oligocene there were no land bridges to Africa and thence South America. From the early Cretaceous to the mid-Pliocene there was no land bridge from North America to South America. An early Cretaceous origin of simians in Laurasia is geographically feasible but it is 90 Ma before the earliest reported simian.

There is fossil support for a Gondwanan origin for simians, which were already diverse in mid to late Eocene in North Africa. An African origin requires spread to both South America and separately to Eurasia. Tertiary land bridges from Africa to Asia are dubious before the late Oligocene, and were intermittent in the Miocene,[18] though connections to Europe existed in the Paleocene.[14] The land bridge to South America was broken about 100 Ma ago. The entry of simians to South America is thus difficult to explain, whether they came from Africa or from Laurasia.

The geographic contrast with the sister group, *Tarsius*, is striking. All alleged tarsiiforms known are from north of the Tethys except for *Afrotarsius*, a single mandible from the Fayum[47] whose tarsiiform status is doubtful.[48]

Omomyiforms were widespread in the northern continents in the Eocene. One European family, the Microchoeridae (including *Pseudoloris*), shows particular affinities to *Tarsius* (see chapter 8). They are now known from China,[49] which was east of the Turgai Sea that dried in the Oligocene. They may have entered Asia via Beringia, as *Shoshonius* from the late early Eocene in North America shows affinities to *Tarsius.*[50] A late date for a land route between Europe and North America could account for this distribution.[47,50-52]

There is thus an interesting geographic pattern among the haplorhines. Simians probably evolved south of the Tethys in Africa, whence they later spread. Tarsiers probably evolved from omomyiforms north of the Tethys in the Eocene or Paleocene. Such models imply that the split between *Tarsius* and the simians was either at the final separation of Laurasia and Gondwanaland (118 Ma ago), or more likely at the time of the Paleocene land bridges (65–56 Ma ago) between Africa and Europe.

The Tethys Sea, which separated Laurasia from North Africa, was open by 118 Ma ago, although land bridges probably existed at about 65 Ma and 56 Ma ago.[14] A vicariant model for euprimate dispersion to or from Laurasia would suggest either an early Cretaceous or a late Cretaceous-Paleocene split between strepsirhines and haplorhines. The later date is supported by the first appearance of Laurasian euprimate fossils in the Eocene. Omomyiforms appear in the later Eocene of the three northern continents more or less simultaneously and adapiforms are known from the lower Eocene in North America and Europe.[14,45] If the strepsirhine/haplorhine split occurred in Laurasia much later (in the Eocene), a dry-land crossing from Eurasia to Africa would not have been possible before the late Oligocene. This is too late to explain the oldest North African fossils.

PLATYRRHINES

The main problem in models of platyrrhine origins is explaining how platyrrhines reached South America before 26 Ma ago. Reconstructions of Caribbean and South Atlantic plate tectonics generally show that South America and North America were separate from the early Cretaceous till the mid-Pliocene, and that South America and Africa separated in the mid-Cretaceous, with a complete land bridge disappearing about 105 Ma ago. By the early Tertiary, when the platyrrhine-catarrhine split may have occurred, the South Atlantic was already wide, and dispersal by this route is problematic even if volcanic islands were present.[3] Estimates of the width of the Mid-Atlantic in the late Eocene range from 1000 to 1750 km; the present width is about 3200 km.[53]

An exactly parallel case is that of hystricognath rodents: two geographic branches of a clade separated by the South Atlantic. Caviomorphs may have entered South America at about the same time as platyrrhines, as they first appear 10 Ma earlier than *Branisella*.[36,54,55]

Some North American paleontologists have long maintained the origin of all anthropoids lay in the Americas. For instance, Gingerich[56] suggested that simians evolved from adapiforms in Asia, with the platyrrhine lineage entering North America via Beringia, thence crossing to South America. Szalay[51] suggested that simians evolved from North American omomyids and later entered South America where the platyrrhines remain to this day. Some then may have crossed to Africa to become catarrhines.

These scenarios have severe problems and can now safely be rejected. First, Szalay's[51] model requires not one but two long ocean crossings by rafting. One such crossing is improbable enough. Second, the presumed directions of ocean currents favor an east-to-west Atlantic crossing and a south-to-north crossing of the proto-Caribbean. In both cases this is the opposite of what these models require. Third, the models suggest that haplorhines may be polyphyletic—a separate origin for *Tarsius*—which most authorities (including Szalay himself) now reject. Gingerich[56] is almost alone in placing *Tarsius* as a sister group to all other living primates.

Fourth, Szalay's[51] model is inconsistent with the earlier dates now accepted for the Fayum primates, as well as the recent discoveries of Paleogene simians in Africa and Asia (unknown when he put his model forward). A west-to-east Atlantic dispersal would predict that haplorhines had a longer fossil history in the Americas than in Africa—the opposite is observed. The downward revision in the ages of the Oligocene Salla fauna has made the African origin model stronger.[57] *Branisella* was once thought to be early Oligocene, but it now dates to 26 Ma ago. The Fayum fauna, once thought to be 25—27 Ma old, now dates to 36 Ma ago.

The largest Cenozoic drop in world mean sea levels apparently occurred about 29 Ma ago (between the dates of Fayum & *Branisella*). Fleagle[57] suggests this was the most likely time for rafting from Africa to occur. This convenient date is now too recent, as the South Atlantic was almost as wide then as it is now. Even if all the continental shelves were exposed, the distance was immense. Ciochon and Chiarelli[53] prefer a late Eocene crossing, consistent with the date of the earliest catarrhines.

It appears that rafting must have occurred at some time in the last 100 Ma. The earlier this is presumed to have occurred the more likely its accuracy, since the distance has been steadily increasing. During the final separation of the South American and African continental shelves between 105 and 85 Ma ago, the region between northeastern South America and West Africa was very active geologically,[3,22] and volcanic islands may have been present.

Madagascar Primates

The Madagascar primates are usually thought to have arrived by rafting across the Mozambique channel from Africa. Madagascar's position relative to Africa has been stable for 120 Ma,[23] and there has been no dry-land connection during that time. An alternative is that primates reached Madagascar from India, to which it was connected until much later. Recent plate reconstructions[9] allow that India might have had a Cretaceous connection to the Horn of Africa as late as 95 Ma ago. If this were the case, and if strepsirhines were a clade by 95 Ma ago, they may have entered Madagascar from Africa via India eliminating the need for any water crossing. However, this scenario requires ancestral strepsirhines in India, for which there is no fossil evidence.

Hominoid Dispersals

The "Neogene Dispersal Phases" between Africa and Asia in the Miocene correlate with the appearance of hominoids in Eurasia, suggesting that dispersal out of Africa occurred only during periods of dry-land connection.[58] Thomas[18] and Steininger[59] both suggest that pliopithecids (or other gibbon ancestors) entered Eurasia during NDP1 (18 ± 1 Ma ago) and that ancestors of *Sivapithecus* followed by *Dryopithecus* dispersed in NDP 2 (15 ± 1 Ma ago).

The pliopithecids are still considered by some to be potential ancestors of gibbons,[60] though others believe them to be the sister group of all living catarrhines.[61,62] In any case, the oldest known hominoid in Asia is the Kamlial molar, dated to 16.1 Ma and possibly a hylobatid. This date is consistent with entry of gibbon ancestors into Asia during the earlier NDP 1. Thus if the Kamlial molar is a hylobatid, paleogeography indicates a 18 ± 1 Ma date for the hylobatid-African ape split. This agrees with the minimum date of 17 to 18 Ma suggested by first appearances of the fossils *Kenyapithecus* and *Afropithecus*, both of which share derived characters with the African hominids.

By contrast, the earliest alleged *Sivapithecus*, a possible orangutan ancestor, is known from Western Asia (Turkey), perhaps from about 14 Ma ago.[63] The oldest known South and East Asian members of the clade are younger, appearing in the Siwaliks at about 12.5 Ma ago. This is consistent with an entry of *Sivapithecus* from Africa into Asia during NDP 2. The simplest explanation for the absence of *Sivapithecus* during the earlier NDP 1 (when pliopithecids dispersed) is that the *Pongo* clade did not then exist. Thus paleogeography suggests a date of 15 ± 1 Ma for the Pongo/African-ape split.

SUMMARY

Rodents, edentates and primates are very divergent placentals. Paleobiogeography supports origins of all three well before the end of the Cretaceous. Within primates, an argument can be made from biogeography that strepsirhines diverged from haplorhines, and possibly tarsiers from simians, also before the end of the Cretaceous. A biogeographical explanation of the origin of platyrrhines in South America is difficult. The presence of apes in Asia (gibbons and orangutan) is explained by a movement out of Africa during the Neogene dispersal phases (NDP1 and 2), during brief periods of contact between Africa and Asia. This would imply divergence times of approximately 18 Ma ago and 15 Ma ago for gibbons and orangutans respectively.

REFERENCES

1. Norrell MA, Novacek MJ. Congruence between superpositional and phylogenetic patterns: comparing cladistic patterns with fossil records. Cladistics 1992; 8:319-337.
2. Irving E. Drift of the major continental blocks since the Devonian. Nature 1977; 270:304-309.
3. Tarling DH. The geologic evolution of South America with special reference to the last 200 million years. In: Ciochon RL, Chiarelli AB, eds. Evolutionary biology of the New World monkeys and continental drift. New York: Plenum Press, 1980:1-41.
4. Audley-Charles MG. Reconstruction of eastern Gondwanaland. Nature 1983; 306:48-50.
5. Veevers JJ, Stagg HMJ, Wilcox JB, Davies HL. Pattern of seafloor spreading (<4mm/year) from breakup (96 Ma) to A20 (44.5 Ma) off the southern margin of Australia. J Aust Geol 1990; 11:499-507.
6. Veevers JJ. Phanerozoic Australia in the changing configuration of Proto-Pangea through Gondwanaland and Pangea to the present dispersed continents. Aust Syst Biol 1991; 4:1-11.
7. Veevers JJ. Phanerozoic earth history in Australia. Oxford: Clarendon, 1984.

8. Scotese CR, Gahagan LM, Larson RL. Plate tectonic reconstructions of the Cretaceous and Cenozoic ocean basins. Tectonophysics 1988; 155:27-48.
9. Weijermars A. Global tectonics since the breakup of Pangaea 180 million years ago: evolution maps and lithographic budget. Earth-Science Reviews 1989; 26:113-162.
10. Kirsch J. Vicariance biogeography. In: Archer M, Clayton G, eds. Vertebrate zoogeography and evolution in Australasia. Western Australia: Hesperian Press, 1984:109-112.
11. Briggs JC. Biogeography and plate tectonics. Amsterdam: Elsevier, 1987.
12. Martin RD. Primate origins and evolution, a phylogenetic reconstruction. London: Chapman and Hall, 1990.
13. White ME. The greening of Gondwana. Balgowlah, New South Wales: Reed, 1986.
14. Gheerbrant E. On the early biogeographical history of the African placentals. Hist Biol 1990; 4:107-116.
15. Fairhead JD. Mesozoic plate tectonic reconstructions of the Central and South Atlantic Ocean: the role of the West and Central African rift system. Tectonophysics 1988; 155:181-191.
16. Fairhead JD, Binks RM. Differential opening of the Central and South Atlantic Oceans and the opening of the West African rift system. Tectonophysics 1991; 187:191-203.
17. Pilbeam D. The descent of hominoids and hominids. Sci Am 1984; 250:60-69.
18. Thomas H. The early and middle Miocene land connection of the Afro-Arabian plate and Asia: a major event for hominoid dispersal? In: Delson E, ed. Ancestors: the hard evidence. New York: Alan Liss, 1985:42-50.
19. Wood AE. The relationships, origin, and dispersal of the hystricognathous rodents. In: Luckett WP, Hartenberger J-L, eds. Evolutionary relationships among rodents: a multidisciplinary analysis. New York: Plenum Press, 1985:475-513.
20. Carroll RL. Vertebrate paleontology and evolution. New York: Freeman, 1988.
21. Thewissen JGM, McKenna MC. Paleobiogeography of Indo-Pakistan: a response to Briggs, Patterson and Owen. Syst Biol 1992; 41:248-251.
22. Mascle J, Blarez E, Marinho M. The shallow structures of the Guinea and Ivory Coast-Ghana transform margins: their bearing on the equatorial Atlantic Mesozoic evolution. Tectonophysics 1988; 155:193-209.
23. Rabinowitz PF, Coffin MF, Falvey D. The separation of Madagascar and Africa. Science 1983; 220:67-69.
24. Metcalfe I. Allochthonous terrance processes in Southeast Asia. Phil Trans R Soc Lond 1990; 331:625-640.
25. Burrett C, Duhig N, Berry R, Varne R. Asian and South-western Pacific continental terranes derived from Gondwana, and their biogeographic significance. Aust Syst Bot 1991; 4:13-24.

26. Brewster B. Antarctica: wilderness at risk. Melbourne: Sun Books, 1982.
27. Archer M. Origins and early radiations of marsupials. In: Archer M, Clayton G, eds. Vertebrate zoogeography and evolution in Australia. Western Australia: Hesperian Press, 1984:585-626.
28. Richardson BJ. A new view of the relationships of Australian and American marsupials. Aust Mammal 1987:71-73.
29. Szalay FS, Novacek MJ, McKenna MC. Mammal phylogeny. New York: Springer-Verlag, 1993.
30. Augee S. Quills and bills: the curious problem of monotreme zoogeography. In: Archer M, Clayton G, eds. Vertebrate zoogeography and evolution in Australasia. Western Australia: Hesperian Press, 1984:567-570.
31. Storch G. *Eomanis waldi*, ein Schuppentier aus dem Mittel-Eozän der "Grube Messel" bei Darmstadt (Mammalia: Pholidota) Fossilfundstelle Messel. Senckenbergiana Lethaea 1978; 59:503-529.
32. Storch G. *Eurotamandua joresi*, ein Myrmecophagide aus dem Eozän der "Grube Messel" bei Darmstadt (Mammalia: Xenarthra). Fossilfundstelle Messel Nr 19. Senckenbergiana Lethaea 1981; 61:247-289.
33. Cifelli RL, Schaff CR, McKenna MC. The relationships of the Arctostylopidae (Mammalia): new data and interpretation. Bull Mus Comp Zool 1989; 152:1-44.
34. De Muizon C, Marshall LG. *Alcidedorbignya inopinata* (Mammalia: Pantodonta) from the early Paleocene of Bolivia: phylogenetic and paleobiogeographic implications. J Paleontol 1992; 66:499-520.
35. Lavocat R. The interrelationships between the African and South American rodents and their bearing on the problem of the origin of the South American monkeys. J Hum Evol 1974; 3:323-326.
36. Wyss AR, Flynn JJ, Norell MA, Swisher CC, Charrier R, et al. South America's earliest rodent and recognition of a new interval of mammalian evolution. Nature 1993; 365:434-437.
37. Graur D, Hide WA, Li WH. Is the guinea pig a rodent? Nature 1991; 351:649-652.
38. Li WH, Hide WA, Graur D. Origin of rodents and guinea-pigs. Nature 1992; 359:277-278.
39. Graur D. Molecular phylogeny and the higher classification of eutherian mammals. Trends Ecol Evol 1993; 8:141-147.
40. Hasegawa M, Cao Y, Adachi J, Yano T. Rodent polyphyly? Nature 1992; 355:595.
41. Honeycutt RL, Adkins RM. Higher level systematics of eutherian mammals: an assessment of molecular characteristics and phylogenetic hypotheses. Annu Rev Ecol Syst 1993; 24:279-305.
42. Cao Y, Adachi J, Yano T, Hasegawa M. Phylogenetic place of guinea pigs: no support of the rodent-polyphyly hypothesis from maximum-likelihood analyses of multiple protein sequences. Mol Biol Evol 1994; 11:593-604.
43. Martin RD. Primate origins: plugging the gaps. Nature 1993; 363:223-234.

44. Hoffstetter R. Phylogeny and geographical deployment of the Primates. J Hum Evol 1974; 3:327-350.
45. Gingerich PD. Early Eocence *Cantius torresi*—oldest primate of modern aspect from North America. Nature 1986; 319:319-321.
46. Klein RG. The human career: human biological and cultural origins. Chicago: University of Chicago Press, 1989.
47. Simons EL, Bown TM. *Afrotarsius chatrathi*, first tarsiiform primate (?Tarsiidae) from Africa. Nature 1985; 313:475-477.
48. Ginsburg L, Mein P. *Tarsius thailandica* nov. sp., premier Tarsiidae (Primates, Mammalia) fossile d'Asie. C R Acad Sc Paris 1987; 304:1213-1215.
49. Beard KC, Qi T, Dawsin MR, Wang B, Li C. A diverse new primate fauna from middle Eocene fissure-fillings in southeastern China. Nature 1994; 368:604-609.
50. Beard KC, Krishtalka L, Stucky RK. First skulls of the early Eocene primate *Shoshonius cooperi* and the anthropoid-tarsier dichotomy. Nature 1991; 349:64-67.
51. Szalay FS. Systematics of the omomyid (Tarsiiformes, Primates) taxonomy, phylogeny and adaptations. Bull Am Mus Nat Hist 1976; 156:157-450.
52. Sigé B, Jaeger J, Sudre J, Vianey-Liaud M. *Altiatlasius kolchii*, n., gen et sp., primate omonyidé du Paléocène supérieur du Maroc, et les origines des euprimates. Palaeontographica Abt A 1990; 214:31-56.
53. Ciochon RL, Chiarelli AB. Paleobiogeographic perspectives on the origin of the Platyrrhini. In: Ciochon RL, Chiarelli AB, eds. Evolutionary biology of the New World monkeys and continental drift. New York: Plenum, 1980:459-493.
54. Fleagle JC. The fossil record of early catarrhine evolution. In: Wood B, Martin L, Andrews P, eds. Major topics in primate and human evolution. Cambridge: Cambridge University Press, 1986:130-149.
55. MacFadden BJ. Chronology of Cenozoic primate localities in South America. J Hum Evol 1990; 19:7-21.
56. Gingerich PD. Eocene Adapidae, paleobiogeography and the origin of South American monkeys. In: Ciochon RL, Chiarelli AB, eds. Evolutionary biology of the New World monkeys and continental drift. New York: Plenum, 1980:123-138.
57. Fleagle JG. Early anthropoid evolution in Africa and South America. In: Else JG, Else PC, eds. Primate evolution. Cambridge: Cambridge University Press, 1986:133-142.
58. Pilbeam D. The descent of hominoids and hominids. Sci Am 1984; 84-96.
59. Steininger FF. Dating the paratethys Miocene hominoid record. In: Else JG, Else PC, eds. Primate evolution. Cambridge: Cambridge University Press, 1986:71-84.
60. Barry JC. A review of the chronology of Sivalik hominoids. In: Else JG, Else PC, eds. Primate evolution. Cambridge: Cambridge University Press, 1986:93-106.
61. Andrews P, Cronin JE. The relationships of *Sivapithecus* and *Ramapithecus*

and the evolution of the orang-utan. Nature 1982; 297:541-545.

62. Delson E. The earliest *Sivapithecus*? Nature 1985; 318:107-108.
63. Andrews P, Tekkaya I. A revision of the Turkish Miocene hominoid *Sivapithecus meteai*. Palaeontology 1980; 23(1):85-95.

CHAPTER 10

THE AGE OF MAMMALS: THE MOLECULAR CLOCK REQUIRES A REASSESSMENT OF DIVERGENCE TIMES

In the previous three chapters we have documented relevant paleontological and biogeographic evidence to show that there is considerable uncertainty in estimates of mammalian divergence times. In chapters 4 and 5 we demonstrated that the rate of DNA evolution is broadly uniform among mammalian lineages. In this chapter we consider the implications for divergence times of molecular evolutionary rate uniformity. There have been many previous studies in which mammalian, and particularly primate, divergence times have been estimated from molecular distances (see for example refs. 1,2,3,4,5). These have almost all involved extrapolation (or interpolation) from one or more assumed divergence times to others. The results, which we do not review in detail here, depend on the validity of assumptions about specific divergence times. We believe that, for the most part, the mammalian fossil record is sufficiently open to varied interpretation that few, if any, divergence times can be definitely assigned from fossil evidence alone. In this chapter we adopt a more cautious approach by considering a range of different evolutionary rates and the implications of each of these. Our conclusion is that a number of mammalian divergence times, both recent and ancient, need to be revised.

DNA DISTANCES AMONG MAMMALIAN TAXA

We have compiled the silent and noncoding nucleotide site distances for several pairs of mammalian taxa whose divergence times range from a few million years to over 100 Ma (Table 10.1). The taxa were selected on the basis that a large sample of nucleotides could be compared between the members of each pair, although the different comparisons did not always involve the same set of genes. The use of different sets of genes to obtain distances for different pairs of taxa could present problems when these distances are compared, since the rate of substitution is known to vary among genomic regions.[6,7] However, where DNA-DNA hybridization data are available (human *vs* chimpanzee, orangutan, Old World monkeys, and New World monkeys; mouse *vs* rat) they provide almost exactly the same distance estimates as the nucleotide sequences. This suggests that the different sets of nucleotide sequences used to obtain the distance estimates are representative of the genome and can thus be compared with each other. The sequences used in the other comparisons also appear representative. The murid–cricetid and the rodent–primate comparisons involved the same set of sequences as the mouse–rat comparison (which was consistent with the DNA–DNA hybridization distances). Also, the set of sequences we used for the placental–marsupial comparisons provide approximately the same estimate of distance between primates and rodents as the sequences we have used for the primate–rodent comparison.

REASSESSING DIVERGENCE TIMES

The divergence-time implications of these distances for a range of possible nucleotide substitution rates are considered in Table 10.1. With regard to primate divergence times it may be useful to refer to Table 8.1. It is clear that none of the rates within the range we have considered is compatible with *all* current estimated divergence times. Rates out-

Table 10.1. Divergence times for mammalian taxa derived from nucleotide sequence differences assuming various constant rates of molecular evolution

	DNA divergence		Divergence time estimates					
			Substitution rate per site per year × 10^9					
Compared taxa	Sequence[1]	DNA-DNA[2] hybridization	1.5	1.75	2.0	2.25	2.5	2.75
Human - Chimp	1.6 ± 0.2	1.7	5.3 ± 0.7	4.6 ± 0.5	4.0 ± 0.5	3.6 ± 0.4	3.2 ± 0.4	2.9 ± 0.4
Human - Orangutan	3.6 ± 0.3	3.7	12 ± 1	10 ± 1	9.0 ± 0.8	8.0 ± 0.7	7.2 ± 0.6	6.5 ± 0.6
Human - Gibbon		4.8	16	14	12	11	9.8	8.9
Human - OWM	7.9 ± 0.4	7.3	26 ± 2	23 ± 1	20 ± 0.8	18 ± 0.7	16 ± 0.6	14 ± 0.6
Human - NWM	13.0 ± 0.7	13.1[3]	43 ± 3	37 ± 2	33 ± 1	29 ± 1	26 ± 1	24 ± 0.9
Human - Rodent	51.6 ± 1.4		172 ± 5	147 ± 4	129 ± 4	115 ± 3	103 ± 3	94 ± 3
Placental - Marsupial	99.0 ± 3.6		330 ±12	283 ±10	248 ± 9	220 ± 8	198 ± 7	180 ± 7
Rat - Mouse	19.9 ± 0.7	19.9	66 ± 3	57 ± 2	50 ± 2	44 ± 2	40 ± 1	36 ± 1
Murid - Cricetid	31.1 ± 0.9		104 ± 3	89 ± 2	78 ± 2	69 ± 2	62 ± 2	57 ± 1

OWM = Old World monkey; NWM = New World monkey

1. Sequence divergence estimates from ref. 8 (Primates); ref. 9 (Human -Rodent, Rat - Mouse, Murid - Cicretid); ref. 10 (Placental - Marsupial).

2. $\Delta T_{50}H$ values from ref. 11 (Primates); ref. 3 (Rat - Mouse)

3. ΔT_m distance from ref. 12

side the range pose even more serious problems. This result mandates some reassessment of divergence times. The question is which ones?

The slowest rate we consider is 1.5×10^{-9} per site per year. This is broadly consistent with current theories of primate divergence times. A human chimpanzee divergence of 5.3 Ma is implied which is well before the first known occurrence of the bipedal *Australopithecus* (the Laetoli footprints at 3.7 Ma) and before the recently discovered 4.4 Ma old *Australopithecus ramidus.*[13] A 12 Ma divergence of humans and orangutans is close to the first definite occurrence of *Sivapithecus* (12.5 Ma), and a 26 Ma divergence of apes and Old World monkeys predates the first occurrence of the earliest presumed ape (*Proconsul*) and the earliest presumed Old World monkeys (*Prohylobates* and *Victoriapithecus*). The problem with this slow rate is its implications for divergence times at the other end of the scale of mammalian evolution. In particular it implies a 330 Ma divergence time for placentals and marsupials. This is three times earlier than most current estimates and, more importantly, it is substantially earlier than current estimates of the divergence of mammals from birds and reptiles (about 300 Ma). It also implies a late Cretaceous (66 Ma) divergence of mice from rats, an early Cretaceous (104 Ma) divergence of murids from cricetids, and a mid Jurassic divergence of rodents from primates. All of these dates currently seem too improbable to consider seriously.

The faster rate of 1.75×10^{-9} per site per year is consistent with a divergence of humans and chimpanzees before *Austalopithecus ramidus* (4.4 Ma). It does, however, imply a 10 Ma divergence of humans and orangutans, which postdates the first occurrence of *Sivapithecus.* This implies that at least part of the range of *Sivapithecus* was not on the orangutan lineage, but was a sister group to all great apes (including humans). The rate implies a marsupial-placental divergence of approximately 283 Ma, implying an even older therian–monotreme divergence and an older still mammalian–bird/reptile divergence. These dates are quite inconsistent with the relatively well documented fossil record of mammal-like reptiles. It also implies a late Jurassic divergence of primates and rodents, a Cretaceous murid–cricetid divergence, and a Paleocene divergence of mice and rats. As was the case for a rate of 1.5×10^{-9} per site per year, these divergence times are vastly greater than those currently accepted, and are probably too early to be seriously considered.

The fastest of the rates we consider in Table 10.1 is 2.75×10^{-9} per site per year. This gives divergence times for placentals and marsupials (180 Ma), murids and crecitids (57 Ma), mice and rats (36 Ma), and rodents and primates (94 Ma) that are closer to the currently accepted times, although they are all still older than these. This rate, however, implies a 24 Ma Old World monkey–New World monkey divergence. This is inconsistent with the occurrence of the fossil primate

Branisella in South America 26 Ma ago. While the affinities of *Branisella* within platyrrhines may be arguable, its location in South America gives a very firm minimum date for the geographic separation of New World monkeys from Old World monkeys. This implies that a rate of 2.75×10^{-9} per site per year can be ruled out.

The next fastest rate, 2.5×10^{-9} per site per year gives a 26 Ma divergence of Old World and New World monkeys and is thus just consistent with *Branisella.* However, it also implies a 16 Ma divergence of apes and Old World monkeys. This substantially postdates *Proconsul,* currently thought to be the first fossil ape, and it also postdates *Kenyapithecus, Afropithecus,* and *Heliopithecus,* all currently thought to be on the great ape lineage, i.e. occurring after the split between great apes and gibbons. More importantly the 16 Ma date is after the occurrence of *Prohylobates* and *Victoriapithecus,* the first known Old World monkeys. These fossils share bilophodont molars with all other Old World monkeys. Bilophodonty is very distinct and its first occurrence probably provides a reliable minimum date for the first occurrence of Old World monkeys. On this basis, a substitution rate of 2.5×10^{-9} seems unlikely.

The two remaining intermediate rates, 2.25×10^{-9} and 2.0×10^{-9} per site per year, both suggest divergence times for all pairs of taxa that are inconsistent with current estimates. They may, nevertheless, be the most acceptable rates because the discrepancies they imply are the least extreme. They imply that the marsupial–placental divergence occurred 220–250 Ma ago. This is much earlier than the 140 Ma that is currently accepted on the basis of the first known occurrence of possible placentals and marsupials.[14] However it is not so early as to be ruled out by the earlier mammal-like reptiles. It is also worth noting that a therian–monotreme split before 200 Ma ago has been suggested previously on the basis of fossil evidence.[15,16] This range of rates also implies an early to mid Cretaceous primate–rodent divergence, a late Cretaceous divergence of cricetids and murids, and an Eocene mouse–rat divergence. All of these are much earlier than currently believed. However, as we showed in chapter 7, none of them is ruled out by the fossil evidence.

A divergence of placental orders during the Cretaceous is strongly indicated by the biogeographical evidence discussed in the previous chapter, and phylogenetic analysis of sequence data (chapter 6) shows that rodents branched early relative to at least some other placental orders. In the absence of any relevant fossil evidence of substance, an early to mid Cretaceous separation of rodents thus seems quite reasonable.

As we discussed in chapter 7, the earliest cricetid fossils are from the Eocene. It is thought that murids diverged from cricetids after this. However, this depends on interpretation of the poor fossil record of a group in which parallel and convergent evolution is known to have occurred with respect to the characters used in making phylogenetic inferences. Current fossil interpretation is not consistent with an Eocene mouse–rat divergence or a late Cretaceous cricetid–murid divergence. However, the interpretation has a weak basis and the earlier dates cannot be ruled out.

The implications of the intermediate range of substitution rates with respect to primate divergences is shown in Figure 10.1. A 29-33 Ma divergence of Old and New World monkeys is implied. This is consistent with the 26 Ma old occurrence of *Branisella* in South America. It does, however, imply that the Fayum primates, which are currently thought to be on the Old World monkey lineage, in fact predate the Old World monkey–New World monkey split. A range of 18 to 20 Ma for the divergence of Old World monkeys is implied. This is consistent with the first evidence of bilophodony, but it implies that *Proconsul*, currently thought to be an ape, is in fact an ancestor or sister group to both apes and Old World monkeys.

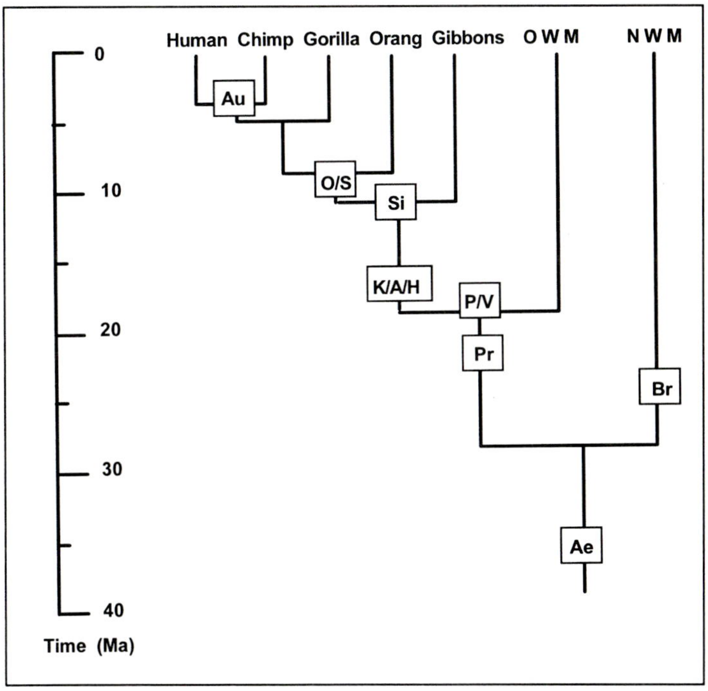

Fig. 10.1. The pattern of simian evolution indicated by a nucleotide substitution rate in the range 2.25×10^{-9} to 2.0×10^{-9} substitutions per site per year. Au: Australopithecus, *Si:* Sivapithecus, *Pr:* Proconsul, *Ae:* Aegyptopithecus, *O:* Ouranopithecus, *S: Samburu mandible, K:* Kenyapithecus, *A:* Afropithecus, *H:* Heliopithecus, *P:* Prohylabates, *V:* Victoriapithecus.

Similarly, an 11 to 12 Ma divergence of gibbons from great apes postdates *Kenyapithecus*, *Afropithecus* and *Heliopithecus*, all of which are currently thought to be on the great ape lineage. These fossils would need to be reinterpreted as ancestors or sister groups to both great apes and gibbons. A divergence of 11 to 12 Ma ago would also postdate the first occurrence of *Sivapithecus* (definitely 12.5 Ma and possibly 14 Ma), implying that at least part of the range of *Sivapithecus* is also an ancestor or sister group to both gibbons and great apes, and is thus definitely not uniquely on the orangutan lineage as some people currently believe. The range of substitution rates we consider likely implies an 8–9 Ma divergence of orangutans from African apes. This would exclude *Sivapithecus* as being on either the African ape or orangutan lineages, and would imply that it is either a sister group or ancestral to both.

If *Sivapithecus* was ancestral to both, then an 11 to 12 Ma split between great apes (including *Homo*) and gibbons has geographical implications, implying that apes evolved for a period outside Africa. From about 16 Ma ago to about 9 Ma ago there are no fossil apes in Africa, although fossil apes may have occurred during this time in Europe, which was connected to Africa, but these have not been reliably dated. In contrast, many ape species, including *Sivapithecus*, occurred in Asia. If the early part of the range of *Sivapithecus* was ancestral to all apes, and the late part occurred on the great ape lineage following separation from gibbons, then that separation would have occurred between two lineages living, as *Sivapithecus* did, in Asia. The geographical separation of orangutans from African apes would then require an explanation of how the common ancestor of African apes and orangutans got from Asia to Africa rather than the other way round. The approximately 9 Ma old Samburu mandible in Africa places a minimum date on the event.

The most dramatic implication of the intermediate range of substitution rates is that it implies a 3.6 to 4 Ma divergence of humans and chimpanzees. Currently a range of dates of 5–7 Ma is widely accepted. This difference is critical in view of the recently discovered *Australopithecus ramidus* at approximately 4.4 Ma. A date of 3.6–4 Ma suggests that *A. ramidus*, rather than being on the human lineage, is in fact a common ancestor or sister group of both humans and chimpanzees. A divergence time in the range of 3.6–4 Ma is also within the known range of *A. afarensis*, suggesting that at least part of the range of *A. afarensis* was also ancestral to both humans and chimpanzees. The earliest definite evidence of bipedalism is the Laetoli footprints, accurately dated to 3.7 Ma. Neither the known remains of *A. ramidus* nor the earliest fossils currently attributed to *A. afarensis* provide conclusive evidence of bipedalism. This range of divergence times does not therefore, necessarily imply that bipedalism is an ancestral trait for both humans and chimpanzees. It does, however, suggest the possibility, and if definite evidence of bipedalism before 4 Ma is discovered it will strongly imply it.

There are two other possible implications of a more recent human–chimpanzee divergence. The first is that between about 4.5 and 3.5 Ma ago five lineages diverged from each other, in other words there appears to have been a star phylogeny. One of these lineages led to *Homo sapiens* and is well represented in the fossil record. Two others, which led to *Australopithecus africanus* and to *Paranthropus* (robust australopithecines, are also well represented in the fossil record but are thought to have became extinct. The remaining two, leading to *Pan* and to *Gorilla* appear to have no fossil representation but survived. The appearance of a star phylogeny generally indicates a lack of data or an inadequacy in interpretation, as we discussed in the context of the divergence of placental orders. We would suggest that the apparent African ape star phylogeny be looked at carefully to see if some other interpretation might not be possible. One possibility that could be considered is that there were only three lineages, the first to separate led to *Paranthropus* and survives as *Gorilla*. The second split led, on the one hand to *H. sapiens* and on the other hand to *A. africanus* which evolved into the surviving chimpanzee species.

The second implication is taxonomic. Historically there has been a progressive increase in the taxonomic proximity of humans to the other African apes. They were once regarded as belonging to separate families (Pongidae and Hominidae). More recently they have been placed in separate tribes (Gorillini, Panini, and Hominini).[17] We suggest that this process be taken further and that both the tribal and generic distinctions be removed. The evolutionary proximity among the African apes, including humans, should be reflected by a similar degree of taxonomic proximity. They should be placed in the same genus—*Homo*. Thus *Pan troglodytes* becomes *Homo troglodytes* and *Pan paniscus* becomes *Homo paniscus*. Gorillas are only slightly more divergent and should also be included in the same genus, thus *Gorilla gorilla* becomes *Homo gorilla*. This revision would imply that the genera *Paranthropus* and *Australopithecus* would also be subsumed in *Homo*.

SUMMARY

In this chapter we have considered the implications of a DNA clock on the divergence times of a range of mammalian taxon pairs that separated at different times. We have avoided any estimate of nucleotide substitution rate based on the assumed divergence time of any particular pair of pairs of taxa. Instead we have considered the divergence-time implications of a range of different substitution rates. Our view is that the most likely rate is in the range of 2.25×10^{-9} to 2.0×10^{-9} per site per year. This is less than previous estimates for the rate in rodents and greater than previous estimates of the rate in primates. It also implies divergence times that we considered that are different from those currently accepted for all the pairs of taxa. Rates either higher or lower than this range, however, imply at least one divergence time that seems to us to be too unlikely to accept. We recognize, however, that this is a matter of judgment, and that we cannot rule out the possibility of either higher or lower rates.

Our purpose here is to point out the alternative implications of the DNA clock. We have drawn attention to the implications we think are most likely. These include a more recent divergence of many primate taxa, including chimpanzees and humans, with the possibility that chimpanzees had bipedal ancestors. We suggest that relevant fossil evidence needs to be re-evaluated by minds open to these implications.

All of the nucleotide substitution rates we have considered imply that the orders of placental mammals and the placental and marsupial sub-classes diverged much earlier than is currently thought. This seems an inescapable conclusion of our analysis. It is interesting to note that a striking parallel situation exists in plants, where molecular data have shown that angiosperms arose substantially earlier than indicated by the fossil record.[18,19]

References

1. Sarich VM, Wilson AC. Immunological time scale for hominid evolution. Science 1967; 158:1200-1203.
2. Hasegawa M, Kishino H, Yano T-A. Dating of the human-ape splitting by a molecular clock of mitochondrial DNA. J Mol Evol 1985; 22:160-174.
3. Catzeflis F, Sheldon FH, Ahlquist JE, Sibley CG. DNA-DNA hybridization evidence of the rapid rate of muroid rodent DNA evolution. Mol Biol Evol 1987; 4:242-253.
4. Li W-H, Tanimura M, Sharp PM. An evaluation of the molecular clock hypothesis using mammalian DNA sequences. J Mol Evol 1987; 25:330-342.
5. Sakoyama Y, Hong K, Byun SM, Hisajima H, Ueda S, et al. Nucleotide sequences of immunoglobulin ε genes of chimpanzee and orangutan: DNA molecular clock and hominoid evolution. Proc Natl Acad Sci (USA) 1987; 84:1080-1084.
6. Wolfe KH, Sharp PM, Li W-H. Mutation rates differ among regions of the mammalian genome. Nature 1989; 337:283-285.
7. Bernardi G, Olofsson B, Filipski J, Zerial M, Salinas J, et al. The mosaic genome of warm-blooded vertebrates. Science 1985; 228:953-958.
8. Easteal S. The relative rate of DNA evolution in primates. Mol Biol Evol 1991; 8:115-127.
9. O'hUigin C, Li W-H. The molecular clock ticks regularly in muroid rodents and hamsters. J Mol Evol 1992; 35:377-384.
10. Easteal S, Collet C. Consistent variation in amino-acid substitution rate, despite uniformity of mutation rate: protein evolution in mammals is not neutral. Mol Biol Evol 1994; 11:643-647.

11. Sibley CG, Alquist JE. DNA hybridization evidence of hominoid phylogeny: Results of an expanded data set. J Mol Evol 1987; 26:99-121.
12. Benveniste RE. The contributions of retroviruses to the study of mammalian evolution. In: McIntyre RJ, ed. Molecular evolutionary genetics. New York: Plenum Press, 1985:359-417.
13. White TD, Suwa G, Asfaw B. *Australopithecus ramidus*, a new species of early hominid from Aramis, Ethiopia. Nature 1994; 371:306-312.
14. Eaton JG. Therian mammals from the Cenomanian (Upper Cretaceous) Dakota formation, southwestern Utah. J Vert Paleontol 1993; 13:105-124.
15. Romer AS. Vertebrate paleontology, 3rd edition. Chicago: University of Chicago Press, 1966.
16. Murray P. Furry egg-layers: the monotreme radiation. In: Archer M, Clayton G, eds. Vertebrate zoogeography and evolution in Australasia. Western Australia: Hesperian Press, 1984:571-584.
17. Groves CP. A theory of human and primate evolution. Oxford: Oxford University Press, 1989.
18. Wolfe KH, Gouy M, Yang Y-W, Sharp PM, Li W-H. Date of the monocot-dicot divergence estimated from chloroplast DNA sequence data. Proc Natl Acad Sci (USA) 1989; 86:6201-6205.
19. Martin W, Gieri A, Saedler H. Molecular evidence for pre-Cretaceous angiosperm origins. Nature 1989; 339:46-48.

CHAPTER 11

Gender Bias: Is Molecular Evolution Male-Driven?

In chapters 3 and 4 we discussed how generation time was first suggested as a factor determining the rate of molecular evolution. It appeared from the results of inter-species protein comparisons and DNA-DNA hybridization experiments that rates of substitution were higher in species with shorter generation times. This was explained as a result of mutations arising as errors in DNA replication, which occurs more frequently in species with short generation times. Such errors might occur either because of a failure of the replication repair of DNA lesions or because of infidelity of the replication process itself. We have pointed out the inadequacy of many of the studies purporting to demonstrate a generation time effect, and we have suggested that in mammals evidence of a generation time effect is lacking. There is, however, another way of assessing the role of DNA replication rate on substitution rate, and that is by comparing the evolutionary rates of X- and Y-chromosomes and autosomes.

In mammals, male germ lines undergo many more rounds of cell division, and hence of DNA replication, than do female germ lines.[1] Thus, if mutation rate is dependent on cell replication time, all else being equal, DNA on Y-chromosomes (which are transmitted entirely in male germ lines) will evolve more rapidly that DNA on autosomes (which spend an equal amount of time in male and female germ lines). Autosomal DNA will, in turn, evolve faster than DNA on X-chromosomes (which spend more time in female than male germ lines).

There have been a number of recent claims either that substitution rates are higher in Y-chromosome DNA, or that they are lower in X-chromosome DNA,[2-5] and that this demonstrates an effect of DNA replication rate on mutation rate. In this chapter we discuss the evidence for this, which is far from convincing.

There are other reasons why this issue is of interest. Crow[6] has raised eugenic implications. He is concerned about a lack of natural selection in current human populations, and suggests that limiting paternity to men under thirty would greatly decrease the load of deleterious mutations. A greater mutation rate in males poses difficulties in understanding the evolution of sexual reproduction. An important advantage of diploid sexual reproduction is the reduced impact of deleterious mutations.[7,8] Redfield[7] has shown that, for females, high male mutation rates outweigh this advantage. Thus, if male mutation rates are as relatively high as is being suggested, it becomes difficult to understand how sexual reproduction has evolved.

HALDANE SUGGESTS A SEX DIFFERENCE IN MUTATION RATES

Haldane[9-11] was first to suggest that mutations might occur more frequently in males than in females due to higher rates of germ-line cell division. The idea arose from an interest in calculating absolute mutation rates in species, particularly humans, in which breeding experiments were impracticable.

Haldane[9] realized that X-linked recessive conditions provide an indirect means of deriving mutation rates, particularly if the conditions have zero fitness (i.e., diseases that involve sterility or death before reproduction). He showed that for an X-linked condition with female and male mutation rates per generation of u and v, respectively, the mean mutation rate per X-chromosome per generation is given by:

$$m = 1/3\ (2u + v), \qquad (11.1)$$

and that when mutation is balanced by selection, for X-linked conditions with incidence x, and fitness among affected males, f,

$$m = 1/3\ (1 - f)\ x. \qquad (11.2)$$

Thus, for X-linked lethals ($f = 0$), 1/3 of all cases result from new mutations, i.e. they are sporadic and the mother of an affected individual is *not* a heterozygote.

Haldane[10] extended this work to directly consider sex differences in mutation rates. As he put it: "if mutation is due to faulty copying of genes at a nuclear division, we might expect it to be commoner in males than females". He showed that the proportion of sons of homozygote normal mothers (i.e. sporadic cases) among hemophiliac males, which is equivalent to the mutation rate, is given by

$$m = (1 - f)\ u\ /\ (2u + v). \qquad (11.3)$$

Using this equation, Haldane was able to estimate u/v (the ratio of female to male mutation rates) for hemophilia, as by 1947 carrier (heterozygote) mothers could usually be detected and hence m could be estimated directly. In several studies of pedigrees of known hemophiliacs Haldane[10,11] found that $u < v$, perhaps by a factor of ten. The ratio of v / u has been called α_m by later workers.

Penrose[12] later pointed out that there was an implied association of risk with the age of the mother's father. He analyzed data relating to three autosomal dominant conditions: tuberous sclerosis, neurofibromatosis (now known to be genetically heterogeneous) and achondroplasia. Only in the latter case was he able to show a significant effect of paternal age. There are a number of other dominant genetic diseases that now have a recognized association with paternal age, including myositis ossificans and Marfan's syndrome.[6,13]

These analyses have limitations that Haldane was aware of. He pointed out that mosaicism in mothers of hemophiliacs might be missed, so leading to overestimates of sporadic cases.[14,15] He also admitted the possibility of ascertainment bias leading to under-estimation of sporadic cases; mild sporadic cases are more likely to be missed in surveys than are families with multiple cases. He also acknowledged the problem of sampling error in that not all heterozygote females will be detected; some will have no sons, and some, by chance, will have no affected sons or maternal grandsons.

In addition to these problems, Haldane's calculations are deficient for a number of reasons that could not have been perceived at the time. First, he and his contemporaries were confusing two separate X-linked diseases with similar and variable phenotypes. Hemophilia is now divided into the more common Hemophilia A (Factor VIII deficiency), and the less common Hemophilia B (Christmas disease or Factor IX deficiency). All cases of Hemophilia A can now be attributed to mutations at the plasma coagulation Factor VIII gene.[16] Hemophilia B results from mutations in the much smaller Factor IX gene.[17] Both genes are on the X-chromosome.

Second, Haldane's indirect method depends on knowing the fitness (or fertility) of affected individuals (and theoretically of heterozygotes, though there is no evidence of heterosis in Hemophilia A or B). This fitness has always been difficult to measure, and has significantly changed in the last 50 years due to better diagnosis and treatment.

Third, Haldane was writing before the molecular nature of genes and mutations was understood. Both Hemophilia A and Hemophilia B are genetically heterogeneous. This is important in the context of the issue that we are concerned with here—the rate of nucleotide substitution in evolution. If cases result from inversions, insertions or deletions rather than from substitutions, then a higher rate of mutation in males does not imply a faster rate of nucleotide substitutions in males.

Recent comparisons of X- and Y-chromosome substitution rates, which purport to demonstrate an effect of DNA replication rate on nucleotide substitution rate,[2-5] have drawn support from the early findings of Haldane, Vogel and Rathenberg[13] and Penrose. It is ironic that, at the same time, characterization of the mutations involved in the diseases investigated by these early workers is leading to the conclusion that DNA replication rate is not an important factor in determining single base mutation rates.

RECENT STUDIES OF HEMOPHILIA A AND B

Analysis of sex differences in rates of mutations responsible for hemophilia resumed in the 1980s. The motivation was now clinical; what was the risk of the mother of a hemophiliac son being a carrier, and hence what was the risk of recurrence? At first, there was little interest in the type of mutation.

Winter et al's[18] results gave α_m = 9.6 (95% confidence limits 2.2 to 41.5), which is consistent with Haldane's earlier result. On the other hand, Rosendahl et al[19] found α_m = 2.1 (95% confidence limits 0.7 to 6.7) and were thus not able to exclude mutation rate equality in males and females. However, their analysis of 6 previous studies (α_m = 1.2-29.3) gave α_m = 3.1 (95% confidence limits 1.4 to 7.1), significantly different from 1, and Bröcker-Vriends et al[20] found α_m = 5.2 (95% confidence limits 1.8 to 15.1), also significantly different from 1 and thus consistent with earlier studies.

As a result of these studies it is now apparent that mutations resulting in Hemophilia A are more frequent in males (as predicted by Haldane) and that risk increases with the age of the maternal grandfather (as predicted by Penrose). However, the relevance of these findings to the issue of substitution rates depend on whether this pattern applies to single nucleotide mutations. For the most part, the type of mutation associated with disease has not been identified, and where it has,[21] the focus has been on other kinds of mutation, such as insertions and deletions. Partly this is because of the complexity of the Factor VIII gene, which makes detecting single-nucleotide mutations laborious. Studies of Factor IX deficiency have been more revealing.

Ludwig et al[17] characterized the causative mutation in 67 cases of Factor IX deficiency. The pedigrees of 40 of these cases were studied, 20 sporadic and 20 with a prior history of Hemophilia B. Many cases were due to insertion or deletion mutations, and overall the results were compatible with α_m = 1, which is inconsistent with the main source of mutations being cell division. In contrast, a similar but smaller study[22] identified the source of mutation as the maternal grandfather in all of 13 Hemophilia B pedigrees, a giving α_m = 11. However, the nature of the 13 mutations[23] is revealing. Of 10 substitutions, 5 were cytosine-to-thymidine transitions.

High rates of C–T transitions in CpG dinucleotides in the Factor IX gene were first noted independently by Koeberl et al[24] and Green et al.[25] Of 22 cases reported by Koeberl et al,[24,26] this type of transition accounted for 38% of all substitutions and had a 24-fold greater rate than other transitions, which occurred 7-fold more frequently than transversions. Transversions at CpG dinucleotides were also elevated 8-fold compared to other transversions.[27] Koeberl et al[28] also found that two C–T transitions at CpG dinucleotides caused arginine-to-nonsense mutations. Such mutations were the cause of about 25% of severe Hemophilia B cases.[25,28]

The significance of the occurrence of mutations at CpG dinucleotides is that these are sites of cytosine methylation. This makes them hotspots for mutations that occur as a result of deamination of 5-methylcytosine.[29] Cytosine methylation is known to occur at CpG dinucleotides in male germ cells but not in female germ cells.[30] This means that any difference in the rate of mutations at CpG dinucleotides between males and females may be due to this difference in methylation rather than any difference in cell generation time.

In the largest study to date, Ketterling et al[31] characterized 25 Factor IX mutations for which the germ line origin had been identified and analyzed these in combination with 18 previously characterized mutations. Of these 43 mutations, 25 occurred in females and 18 in males. The apparent female-origin excess may be due to male origins being undersampled, for example in great-grandfathers who are rarely available for testing. No sex difference was evident in the rate of occurrence of deletions, but there was a 3.5-fold predominance of single-nucleotide mutations in males. However, this is entirely accounted for by an 11-fold difference in the rate of transitions at CpG dinucleotides.

In summary, molecular genetic studies of hemophilia have confirmed Haldane's and Penrose's predictions of higher mutations in males, but they do not provide support for the supposed mechanism—a lower germ-cell generation time. There does appear to be an excess of mutations in males in Hemophilia A, but the nature of the mutations involved in this difference is not known. In Hemophilia B, the greater rate of single base mutations in males is better explained by methylation of male germ cells rather than by lower cell generation time. This interpretation is consistent with the evidence from relative rate tests that the rate of molecular evolution does not vary in any systematic way among species with different generation times. Thus, contrary to the suggestion of Li,[32] these recent molecular genetic studies do not provide support for a generation-time effect on the rate of substitution of single base mutations.

MALE DRIVEN EVOLUTION?

Another approach to investigating differences in mutation (or substitution) rate between male and female germ lines is to compare the rates of substitution of sequences on X- and Y-chromosomes and on autosomes. Since a Y-chromosome spends all its time in males, if mutations arise from errors in DNA replication, the expected mutation rate per generation for Y-chromosomes is given by α, the male-to-female ratio of germ cell divisions. Similarly, since an autosome spends half its time in males and half its time in females, the mean expected autosomal mutation rate per generation is $1/2 \times \alpha + 1/2 \times 1 = (1 + \alpha)/2$. In the same way, since an X-chromosome spends 2/3 of its time in females and only one third in males, the mean expected X-chromosome mutation rate per generation is $1/3 \times \alpha + 2/3 \times 1 = (2 + \alpha)/3$.[4,5]

When $\alpha >> 1$, the relative mutation frequencies will be: $R_{X/A} = 2/3$; $R_{Y/A} = 2$; $R_{Y/X} = 3$, where X, Y, and A are the mutation frequencies for X-, and Y-chromosomes and for autosomes respectively.[4,5]

To test whether germ-line sex differences affect mutation rates, Miyata et al[4,5] compared the sequences of 41 homologous genes in humans and rodents, of which six were X-linked. For silent sites the rate of substitution in the six X-linked genes was 0.58 times that of the autosomal genes, which approached the predicted 2/3 for $\alpha >> 1$. The problem with this analysis is that the autosomal and X-linked genes being compared were not related either functionally or evolutionarily. The possibility that factors associated with sequence composition or context accounted for the mean differences in substitution rate could not be ruled out. Miyata et al[5] also compared the sequences of two reverse-transcribed arginosuccinate pseudogenes, one on the Y-chromosome and one on an autosome, relative to the sequence of the arginosuccinate gene. They found that, as predicted, the Y-linked pseudogene had evolved approximately twice as fast as the autosomal pseudogene. However, once again the possibility of factors other than cell generation time accounting for the difference could not be ruled out.

In mammals, male sex is determined by a gene that lies on the distal short arm of the Y-chromosome near the pseudoautosomal region. The first candidate gene identified[33] coded for a protein with a domain consisting of 13 zinc fingers, referred to as *ZFY*. The *ZFY* gene turned out not to be involved in sex determination. This was indicated by the presence of conserved paralogous genes on X-chromosomes (*ZFX*)[34] and the autosomal location of the homologous genes in marsupials[35] and birds.[36] Investigation of the evolution of the ZFY and ZFX genes and their homologues in other mammals has, however, been important in the debate about α_m.

Lanfear and Holland[36] compared partial sequences of *ZFY*-related genes on the X- and Y-chromosomes of various mammals and birds. From comparisons of *ZFX* with mouse *Zfx* and of *ZFY* with mouse *Zfy* they estimated $R_{Y/X} = 2$, which is less than the value of 3 predicted when $\alpha >> 1$. This analysis assumes that *ZFX* and *ZFY* are orthologous with *Zfx* and *Zfy* respectively, and that no recombination or gene conversion has occurred between the X- and Y-linked genes.

Hayashida et al[37] suggested that neither of these assumptions is valid. The occurrence of both Y-chromosome and X-chromosome genes in a variety of placental orders but not in marsupials suggests that the genes diverged after the divergence of placentals from marsupials but before the divergence of placental orders. If this were the case there would be a closer relationship among X-chromosome genes in species from different orders than between X-chromosome and Y-chromosome genes in the same order. For the sequences analyzed by Lanfear and Holland[36] the human *ZFX* and *ZFY* genes are more closely related to

QUESTIONNAIRE

Receive a FREE BOOK of your choice

Please help us out—Just answer the questions below, then select the book of your choice from the list on the back and return this card.

R.G. Landes Company publishes five book series: *Medical Intelligence Unit, Molecular Biology Intelligence Unit, Neuroscience Intelligence Unit, Tissue Engineering Intelligence Unit* and *Biotechnology Intelligence Unit.* We also publish comprehensive, shorter than book-length reports on well-circumscribed topics in molecular biology and medicine. The authors of our books and reports are acknowledged leaders in their fields and the topics are unique. Almost without exception, there are no other comprehensive publications on these topics.

Our goal is to publish material in important and rapidly changing areas of bioscience for sophisticated scientists. To achieve this goal, we have accelerated our publishing program to conform to the fast pace in which information grows in bioscience. Most of our books and reports are published within 90 to 120 days of receipt of the manuscript.

Please circle your response to the questions below.

1. We would like to sell our *books* to scientists and students at a deep discount. But we can only do this as part of a prepaid subscription program. The retail price range for our books is \$59-\$99. Would you pay \$196 to select four *books* per year from any of our Intelligence Units–\$49 per book–as part of a prepaid program?

   Yes **No**

2. We would like to sell our *reports* to scientists and students at a deep discount. But we can only do this as part of a prepaid subscription program. The retail price range for our reports is \$39-\$59. Would you pay \$145 to select five *reports* per year–\$29 per report–as part of a prepaid program?

   Yes **No**

3. Would you pay \$39–the retail price range of our books is \$59-\$99–to receive any single book in our Intelligence Units if it is spiral bound, but in every other way identical to the more expensive hardcover version?

   Yes **No**

To receive your free book, please fill out the shipping information below, select your free book choice from the list on the back of this survey and mail this card to:

R.G. Landes Company, 909 S. Pine Street, Georgetown, Texas 78626 U.S.A.

Your Name ____

Address ____

City ____________________ State/Province: ____________________

Country: ____________________ Postal Code: ____________________

My computer type is Macintosh______ ; IBM-compatible _____ ; Other _____

Do you own ____ or plan to purchase ___ a CD-ROM drive?

Available Free Titles

Please check three titles in order of preference.
Your request will be filled based on availability. Thank you.

- ❐ Water Channels
  Alan Verkman,
  University of California-San Francisco
- ❐ The Na,K-ATPase: Structure-Function Relationship
  J.-D. Horisberger, University of Lausanne
- ❐ Intrathymic Development of T Cells
  J. Nikolic-Zugic,
  Memorial Sloan-Kettering Cancer Center
- ❐ Cyclic GMP
  Thomas Lincoln, University of Alabama
- ❐ Primordial VRM System and the Evolution of Vertebrate Immunity
  John Stewart, Institut Pasteur-Paris
- ❐ Thyroid Hormone Regulation of Gene Expression
  Graham R. Williams, University of Birmingham
- ❐ Mechanisms of Immunological Self Tolerance
  Guido Kroemer, CNRS Génétique Moléculaire et Biologie du Développement-Villejuif
- ❐ The Costimulatory Pathway for T Cell Responses
  Yang Liu, New York University
- ❐ Molecular Genetics of Drosophila Oogenesis
  Paul F. Lasko, McGill University
- ❐ Mechanism of Steroid Hormone Regulation of Gene Transcription
  M.-J. Tsai & Bert W. O'Malley, Baylor University
- ❐ Liver Gene Expression
  François Tronche & Moshe Yaniv,
  Institut Pasteur-Paris
- ❐ RNA Polymerase III Transcription
  R.J. White, University of Cambridge
- ❐ src Family of Tyrosine Kinases in Leukocytes
  Tomas Mustelin, La Jolla Institute
- ❐ MHC Antigens and NK Cells
  Rafael Solana & Jose Peña,
  University of Córdoba
- ❐ Kinetic Modeling of Gene Expression
  James L. Hargrove, University of Georgia
- ❐ PCR and the Analysis of the T Cell Receptor Repertoire
  Jorge Oksenberg, Michael Panzara & Lawrence Steinman, Stanford University
- ❐ Myointimal Hyperplasia
  Philip Dobrin, Loyola University
- ❐ Transgenic Mice as an In Vivo Model of Self-Reactivity
  David Ferrick & Lisa DiMolfetto-Landon,
  University of California-Davis and Pamela Ohashi,
  Ontario Cancer Institute
- ❐ Cytogenetics of Bone and Soft Tissue Tumors
  Avery A. Sandberg, Genetrix & Julia A. Bridge,
  University of Nebraska
- ❐ The Th1-Th2 Paradigm and Transplantation
  Robin Lowry, Emory University
- ❐ Phagocyte Production and Function Following Thermal Injury
  Verlyn Peterson & Daniel R. Ambruso,
  University of Colorado
- ❐ Human T Lymphocyte Activation Deficiencies
  José Regueiro, Carlos Rodríguez-Gallego and Antonio Arnaiz-Villena,
  Hospital 12 de Octubre-Madrid
- ❐ Monoclonal Antibody in Detection and Treatment of Colon Cancer
  Edward W. Martin, Jr., Ohio State University
- ❐ Enteric Physiology of the Transplanted Intestine
  Michael Sarr & Nadey S. Hakim, Mayo Clinic
- ❐ Artificial Chordae in Mitral Valve Surgery
  Claudio Zussa, S. Maria dei Battuti Hospital-Treviso
- ❐ Injury and Tumor Implantation
  Satya Murthy & Edward Scanlon,
  Northwestern University
- ❐ Support of the Acutely Failing Liver
  A.A. Demetriou, Cedars-Sinai
- ❐ Reactive Metabolites of Oxygen and Nitrogen in Biology and Medicine
  Matthew Grisham, Louisiana State-Shreveport
- ❐ Biology of Lung Cancer
  Adi Gazdar & Paul Carbone,
  Southwestern Medical Center
- ❐ Quantitative Measurement of Venous Incompetence
  Paul S. van Bemmelen, Southern Illinois University and John J. Bergan, Scripps Memorial Hospital
- ❐ Adhesion Molecules in Organ Transplants
  Gustav Steinhoff, University of Kiel
- ❐ Purging in Bone Marrow Transplantation
  Subhash C. Gulati,
  Memorial Sloan-Kettering Cancer Center
- ❐ Trauma 2000: Strategies for the New Millennium
  David J. Dries & Richard L. Gamelli,
  Loyola University

each other than either is to the corresponding gene in mice. Hayashida et al[37] explain this discrepancy as resulting from a gene conversion of *ZFY* by *ZFX*. This pattern of relationship invalidates Lanfear and Holland's[36] conclusions.

Using the same data set, however, Hayashida et al[37] obtained an estimate of $R_{Y/X}$ (not $R_{Y/A}$ as they suggest) = 1.2, by comparing *ZFX* and *ZFY* relative to the outgroup mouse *Zfx*. They suggest that this is "qualitatively" consistent with prediction. It is, however, substantially less than the value of three expected when $\alpha >> 1$, and may not be significantly different from one, although no statistical test results were presented. Pamilo and Bianchi[38] provided further evidence of gene conversion between *Zfy* and *Zfx* by comparing the phylogenies of the 5' and 3' sections of the genes. They also suggest that $R_{Y/X} \approx 2$, although this is based on sequence comparisons made in the context of phylogenies of arbitrary 5' and 3' divisions of the gene, and its validity is hard to assess.

Shimmin et al[2] compared *ZFY* and *ZFX* intronic sequences among humans, orangutans, baboons and squirrel monkeys. They found that $R_{Y/X} \approx 2.3$, implying that $\alpha_m \approx 6$ (95% confidence limits 2 to 84). Their analysis appears not to be confounded by the occurrence of gene conversion. They comment on the fact that their result implies that, although the male mutation rate is higher than the female rate, it is not as high as might be expected given the difference in cell generation time and that this suggests a role for replication-independent mutagenic factors.

Chang et al[3] extended this analysis to rodents by comparing intronic sequences of *Zfx* and *Zfy* in mice and rats. They estimate $R_{Y/X} \approx 1.42$, implying $\alpha_m \approx 1.8$ (95% confidence interval 1 to 3, thus not significantly different from 1). They suggest that this lower value of α_m in rodents compared with primates is due to shorter generation time which results in a smaller difference in the number of cell divisions in male and female lineages.

All these studies indicate a greater rate of substitution at the *ZFY*-related genes in male compared with the female lineages, although only in the primate comparisons is this result unequivocal. However, the rate is not as great as would be expected if $\alpha >> 1$ and if the mutations substituted in evolution arise as errors in DNA replication. Chang et al[3] suggest that this may be because, at least in rodents, α is not as great as has been supposed. The alternative is that single base mutations do not arise as errors in DNA replication.

There are a number of ways of explaining the difference in substitution rates between the X- and Y-linked genes other than in terms of cell generation times. The difference may reflect the occurrence of cytosine methylation in male germ cells but not in female germ cells. As discussed above, this appears to be the explanation of the male–female difference in mutation rate in the coagulation Factor IX gene. This possibility could be investigated by comparing mutation rates at CpG dinucleotides, although identifying these may be difficult as they are hypermutable and thus often have only transitory existence.

Another possible explanation for the difference in substitution rate is that, although the substitutions investigated have been at either silent or noncoding sites, a proportion of them involve the substitution of slightly deleterious mutations. We discussed in chapter 4 how slightly deleterious mutations are more readily substituted in small populations. The effective population size for Y-chromosomes, which are haploid and occur only in males, is thus 1/4 that of autosomes and 1/3 that of X-chromosomes. Thus substitution rates of slightly deleterious mutations occurring on Y-chromosomes will be greater than those occurring on X-chromosomes or autosomes in the same population.

Whatever the explanation for the difference in substitution rates between *ZFY* and *ZFX* in primates, the difference does not appear to be typical of the X- and Y- chromosomes as a whole. Ellis et al[39] compared sequences on either side of the pseudoautosomal boundary in four great apes (including humans) and two Old World monkey species. They concluded that the substitution rate is in the order Y-chromosome > pseudoautosome > X-chromosome, but the differences are very small. Comparison between ape and Old World monkey sequences give the following mean percent sequence differences: Y-chromosome 10.55; autosome 9.47; X-chromosome 9.18. These give: $R_{Y/X} = 1.15$; $R_{X/A} = 0.97$; and $R_{Y/A} = 1.11$ (implying $\alpha_m = 1.25$). Thus while there may be some chromosome effect on substitution rate in this region, the effect is very small compared with the intronic sequences of the *ZFY*-related genes and much smaller than predicted if single base mutations arise predominantly as errors in DNA replication.

SUMMARY

There appears to be a higher rate of single nucleotide mutations leading to coagulation Factor IX deficiency in male lineages than in female lineages. This is a result of differences in mutation rate at CpG dinucleotides and is explained by the occurrence of cytosine methylation in male but not in female germ lines. It is not a reflection of the difference in cell generation time. Estimation of substitution rates from comparative sequence analysis suggests a higher rate of substitution in male germ lines, although the evidence is not consistent and far from conclusive.

Even if this difference is confirmed it does not necessarily imply that cell generation time has an effect of substitution rate. The rate of cell division is not the only way in which male and female lineages differ. A substitution rate difference may be explained by the differential occurrence of methylation or by an the increased substitution rate of slightly deleterious mutations due to a smaller effective population size for Y-chromosomes. More extensive sequence comparisons and more careful analysis of data are required before any definite conclusions can be reached on this issue.

References

1. Vogel F, Kopun M, Rathenberg R. Mutation and molecular evolution. In: Goodman M, Tashian, Tashian JH, eds. Molecular anthropology. New York: Plenum Press, 1994:13-33.
2. Shimmin LC, Chang BH-J, Li W-H. Male-driven evolution of DNA sequences. Nature 1993; 362:745-747.
3. Chang BH-J, Shimmin LC, Shyue S, Hewett-Emmett D, Li W H. Weak male-driven molecular evolution in rodents. Proc Natl Acad Sci (USA) 1994; 91:827-831.
4. Miyata T, Hayashida H, Kuma K, Mitsuyasu K, Yasunga T. Male-driven molecular evolution: a model and nucleotide sequence analysis. Cold Spring Harbor Symp Quant Biol 1987; 52:863-867.
5. Miyata T, Kuma K, Iwabe N, Hayashida H, Yasunaga T. Different rates of evolution of autosome-, X-chromosome- and Y-chromosome- linked genes: hypothesis of male-driven molecular evolution. In: Takahata N, Crow JF, eds. Population biology of genes and molecules. Tokyo: Baifukan, 1990:341-357.
6. Crow JF. How much do we know about spontaneous human mutation rates. Envir Mol Mutagen 1993; 21:122-129.
7. Redfield RJ. Male mutation rates and the cost of sex for females. Nature 1994; 369:145-147.
8. Kondrashov AS. Sex and deleterious mutation. Nature 1994; 369:99-100.
9. Haldane JBS. The rate of spontaneous mutation of a human gene. J Genetics 1935; 31:317-326.
10. Haldane JBS. The mutation rate of the gene for hemophilia, and its segregation ratios in males and females. Annal Hum Genet 1946; 13:262-272.
11. Haldane JBS. The formal genetics of man. Proc Roy Soc Lond 1948; 135:147-170.
12. Penrose LS. Parental age and mutation. The Lancet 1955; 312-313.
13. Vogel F, Rathenberg R. Spontaneous mutation in man. In: Harris H, Hirschhorn K, eds. Advances in human genetics vol.5. New York: Plenum Press, 1975:223-318.
14. Taylor SAM, Deugau KV, Lillicrap DP. Somatic mosaicism and female-to-female transmission in a kindred with hemophilia B (Factor IX deficiency). Proc Natl Acad Sci (USA) 1991; 88:39-42.
15. Jeanpierre M. Germinal mosaicism and risk calculation in X-linked diseases. Am J Hum Genet 1992; 60:960-967.
16. Naylor JA, Green PM, Rizza CR, Giannelli F. Factor VIII gene explains all cases of hemophilia A. The Lancet 1992; 340:1066-1067.
17. Ludwig M, Grimm T, Brackmann HH, Olek K. Parental origin of Fac-

tor IX gene mutations, and their distribution in the gene. Am J Hum Genet 1992; 50:164-173.
18. Winter RM, Tuddenham EGD, Goldman E, Matthews KB. A maximum likelihood estimate of the sex ratio of mutation rates in hemophilia A. Hum Genet 1983; 64:156-159.
19. Rosendaal FR, Bröcker-Vriends AHJT, van Houwelingen JC, Smit C, Varekamp I, et al. Sex ratio of the mutation frequencies in hemophilia A: estimation and meta-analysis. Hum Genet 1990; 86:139-146.
20. Bröcker-Vriends AHJT, Rosendaal FR, van Houwelingen JC, Bakker E, van Ommen GJB, et al. Sex ratio of the mutation frequencies in hemophilia A: coagulation assays and RFLP analysis. J Med Genet 1991; 28:672-680.
21. Rossiter JP, Young M, Kimberland ML, Hutter P, Ketterling RP, et al. Factor VIII gene inversions causing severe hemophilia A originate almost exclusively in male germ cells. Hum Mol Genet 1994; 3:1035-1039.
22. Montandon AJ, Green PM, Bentley DR, Ljung R, Kling S, et al. Direct estimate of the hemophilia B (Factor IX deficiency) mutation rate and of the ratio of the sex-specific mutation rates in Sweden. Hum Genet 1992; 89:319-322.
23. Gianneli F, Saad S, Montandon AJ, Bentley DR, Green PM. A new strategy for the genetic counselling of diseases of marked mutational heterogeneity: hemophilia B as a model. J Med Genet 1992; 29:602-607.
24. Koeberl DD, Bottema CDK, Buerstedde J, Sommer SS. Functionally important regions of the Factor IX gene have a low rate of polymorphism and a high rate of mutation in the dinucleotide CpG. Am J Hum Genet 1989; 45:448-457.
25. Green PM, Bentley DR, Mibashan RS, Nilsson IM, Giannelli F. Molecular pathology of hemophilia B. EMBO J. 1989; 8:1067-1072.
26. Koeberl DD, Bottema CDK, Sarkar G, Ketterling RP, Chen S-H, et al. Recurrent nonsense mutations at arginine residues cause severe hemophilia B in unrelated hemophiliacs. Hum Genet 1990; 84:387-390.
27. Sommer SS. Assessing the underlying pattern of human germline mutations: lessons from the Factor IX gene. FASEB 1992; 6:2767-2774.
28. Koeberl DD, Bottema CDK, Ketterling RP, Bridge PJ, Lillicrap DP, et al. Mutations causing hemophbilia B: Direct estimate of the underlying rates of spontaneous germ-line transitions, transversions, and deletions in a human gene. Am J Hum Genet 1990; 47:202-217.
29. Coulondre C, Miller JH, Farabaugh PJ, Gilbert W. Molecular basis of base substitution hotspots in *Escherichia coli.* Nature 1978; 274:775-780.
30. Driscoll DJ, Migeon BR. Sex difference in methylation of single-copy genes in human meiotic germ cells: implications for X-chromosome inactivation, parental imprinting, and origin of CpG mutations. Somat Cell Mol Genet 1990; 16:267-282.
31. Ketterling RP, Vielhaber E, Bottema CDK, Schaid DJ, Cohen MP, et al. Germ-line origins of mutation in families with hemophilia B: the sex ratio varies with the type of mutation. Am J Hum Genet 1993; 52:152-166.

32. Li W-H. So, what about the molecular clock hypothesis? Curr Op Genet Devel 1993; 3:896-901.
33. Page DC, Mosher R, Simpson EM, Fisher EMC, Mardon G, et al. The sex-determining region of the human Y-chromosome encodes a finger protein. Cell 1987; 51:1091-1104.
34. Schneider-Gadicke A, Beer-Romero P, Brown LG, Nussabaum R, Page DC. *ZFX* has a gene structure similar to *ZFY*, the putative human sex determinant, and escapes X inactivation. Cell 1989; 57:1247-1258.
35. Sinclair AH, Foster JW, Spencer JA, Page DC, Palmer M, et al. Sequences homologous to *ZFY*, a candidate human sex-determining gene, are autosomal in marsupials. Nature 1988; 336:780-783.
36. Lanfear J, Holland PWH. The molecular evolution of *ZFY*-related genes in birds and mammals. J Mol Evol 1991; 32:310-315.
37. Hayashida H, Kuma K, Miyata T. Interchromosomal gene conversion as a possible mechanism for explaining divergence patterns of *ZFY*-related genes. J Mol Evol 1992; 35:181-183.
38. Pamilo P, Bianchi NO. Evolution of the *Zfx* and *Zfy* genes: rates and interdependence between the genes. Mol Biol Evol 1993; 10:271-281.
39. Ellis N, Yen P, Neiswanger K, Shapiro LJ, Goodfellow PN. Evolution of the pseudoautosomal boundary in Old World monkeys and great apes. Cell 1990; 63:977-986.

CHAPTER 12

The Mouse and the Turtle: Does Metabolic Rate Affect Substitution Rate?

Generation time, or cell generation time, has been the factor mainly considered as potentially important in affecting substitution rates among lineages. However, it has also been suggested that metabolic rate may affect mutation rate and hence substitution rate.[1] This suggestion is based on two kinds of evidence. The first comes from experimental work showing that metabolic rate affects the rate of DNA damage. The second is the indication that rates of nucleotide substitution in the mitochondrial genomes of ectothermic species such as sharks,[2] fish and frogs,[3] and turtles[4,5] are slower than the rates in endothermic mammals. If metabolic rate does affect substitution rate this has important implications for the molecular clock, as it implies different substitution rates in species with different metabolic rates. The evidence we have presented thus far suggests that in mammals this is not the case. In this chapter we critically evaluate the arguments that implicate metabolic rate as a factor affecting mutation rate.

OXIDATIVE DNA DAMAGE AND MUTAGENESIS

Endogenous oxygen free radicals, produced as by-products of cellular processes such as respiration, phagocytosis and cell injury, cause damage to DNA, RNA, proteins and lipid membranes; this is despite efficient mechanisms for their removal. This damage is thought to cause various degenerative diseases including cancer, cardiovascular disease, and brain dysfunction, as well as being associated generally with the process of aging.[6-9] The free radicals that most likely interact with nucleotides in DNA and RNA are superoxide (O_2^-), hydrogen peroxide (H_2O_2) and the hydroxyl ion (OH^-). Oxidated nucleotides are normally excised by a DNA polymerase during the repair process, hydrolyzed to deoxynucleosides and eventually secreted in the urine. Although the mechanisms for the repair of DNA damage are highly efficient, they can result in the incorporation of an incorrect base, i.e. a mutation. Mutations can also arise from a failure to repair modified bases which cause base changes during subsequent DNA replication.

There are some 20 modified bases found in urine that are thought to be the by-products of the excision-repair process acting on DNA damaged by oxidation.[10] It is the differences in concentration of some of these bases in the urine of different mammalian species that gave rise to the theory that metabolic rate affects mutation rate.[11,12] It should be stressed that mutation rate is not being assayed directly here. The assay is of excreted levels of DNA repair by-products, which are presumed to correlate with the degree of DNA damage. A relationship between these levels and mutation rates has never actually been demonstrated. In fact it has not been unequivocally demonstrated the urinary deoxynucleosides are *solely* the product of DNA repair of deoxynucleotides already incorporated into chromosomal DNA.[13,14] They may, in part, be the product of modification of free deoxynucleosides.

Adelman et al[14] reported the concentrations of thymine glycol (5,6-dihydro-5,6-dihydroxythymine) and thymidine glycol in the urine of humans, monkeys, rats and mice (Fig. 12.1B, C). The urinary outputs of thymine glycol from rats and mice are very similar and both are significantly higher (5- to 13-fold higher) than those of humans and monkeys. The measured levels of urinary output of thymidine glycol from monkeys, rats and mice, on the other hand, overlap to a large degree. The range of values obtained from humans is not clear in this case, but appears lower than the other three species.

Shigenaga et al[15] analyzed the concentration of the modified nucleoside 8-hydroxy-2'-deoxyguanosine in the urine of humans, rats and mice. Their results, expressed as a function of metabolic rate, measured as oxygen consumption, for each species, are presented in Figure 12.1A. The results are inconclusive. Rats, which are estimated to have a metabolic rate 5.6-fold higher than humans, do not have a significantly higher urinary output of the modified nucleoside. Mice, on the other hand, with a metabolic rate 10.6-fold and 1.9-fold higher than humans and rats, respectively, have a significantly greater level of excreted 8-hydroxy-2'-deoxyguanosine than that in either rats (2-fold higher) or humans (3-fold higher).

The failure to obtain consistent trends in the urinary secretion of the different modified deoxynucleosides underscores the problems of measuring oxidative DNA damage in this indirect fashion, and the difficulties of interpreting inter-species comparisons. Clearly, the choice of modified nucleoside used to monitor the degree of oxidative damage is critical. There are other factors that also complicate the interpretation of the inter-species comparisons.

Fraga et al[16] showed that, in rats, the urinary output of oxidatively-modified bases, in particular 8-hydroxy-2'-deoxyguanosine, may decrease with age, by as much as 3-fold. The decrease in urinary concentration correlates with an increased concentration in the DNA of specific organs that are metabolically very active (e.g. liver, kidney and intestine). This contrasts preliminary evidence which suggests an age-independent excretion of thymine glycol and thymidine glycol in the urine of humans over the age-span of 20 to 84 years.[6] There are thus differences in the effect of age on the metabolism of different modified bases. There may also be differences in the effect of age among species. It is certainly important in inter-species comparisons that the compared individuals be matched for age relative to the total life-span of the species. This has not been done. In the interspecific comparisons described above, individuals from different species were from various age groups, and in different experiments different age groups were used and the results directly compared.

There is also evidence of variation in excretion rates within individuals over short time periods. Park et al[17] examined the degree of reproducibility of an assay for 8-oxo-7,8-dihydro-2'-deoxyguanosine, another by-product of oxidative damage to DNA, and the degree of consistency of excretion levels within individual rats at different times. Despite their assay being more precise than that used by Shigenaga et al,[15] they found that variation between assays of the same sample was as high as 20%. They also found that excretion rates could vary as much as 2-fold in the same individual sampled 130 days apart.

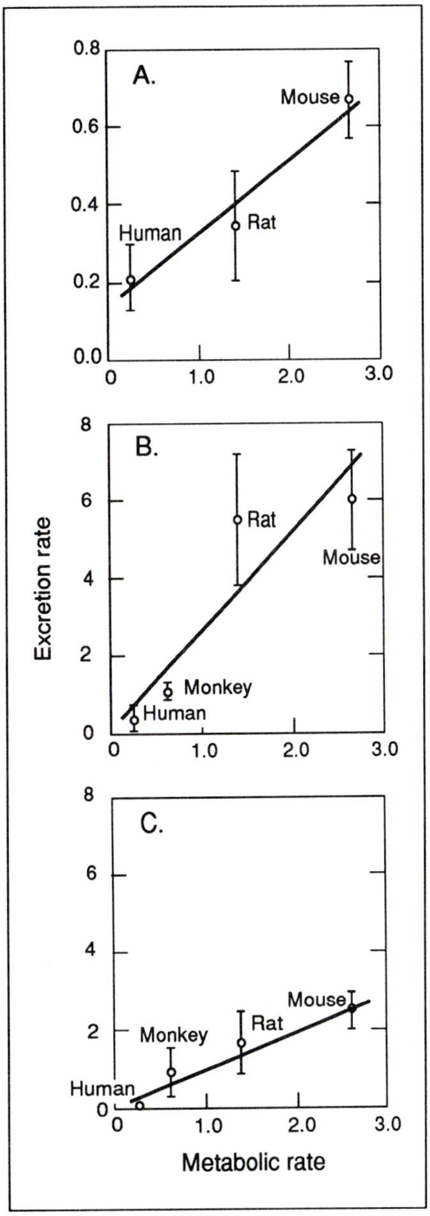

Fig. 12.1. Excretion rates of modified nucleosides in mammal species as a function of metabolic rate. The urinary output (nmol/kg/day) 8-hydroxy-2'-deoxyguanosine (A), thymine glycol (B), and thymidine glycol (C) were measured and plotted against oxygen consumption (O_2 mg/g/hour). Modified and reproduced with permission from M K Shigenaga et al, Proc Natl Acad Sci (USA) 1989; 86:9697-9701 and R Adelman et al, Proc Natl Acad Sci (USA) 1988; 85:2706-2708.

In addition they found that among 20 individuals there was a 6-fold range in the level of 8-oxo-7,8-dihydro-2'-deoxyguanosine secreted in the urine. A similar study on five individuals showed a 6-fold range in levels of thymine glycol excreted in a 24-hour period,[13] although excreted levels of thymidine glycol over the same 24-hour period were more consistent. Park et al[17] also showed that diet affects, to varying degrees, both the mean and the variance of urinary output of the different nucleoside derivatives of 8-oxoguanine.

The finding of such extensive variation within and among individuals is perhaps not surprising. Cells possess multiple DNA repair mechanisms and metabolic pathways for the processing of the different oxidated deoxynucleotides and deoxynucleosides. In addition, there are differences in the ways in which the various oxygen free radicals are produced in the first place,[7] and in the ways in which they act. The variation does, however, make it difficult to draw any strong conclusions from the cross-species comparisons that have been made to date. The variation also underscores the complexity of the relationship of the assay (amount of DNA-repair byproducts in urine) to the thing being assayed (oxidative damage to DNA), let alone its relationship to mutation rate. This complexity means that even the most carefully designed use of this kind of assay in interspecies comparisons will not give conclusive results with respect to the effect of metabolic rate on mutation rate.

In the studies discussed above the assays are designed to estimate levels of oxidative damage (as potential sources of mutation) to DNA in somatic rather than germ cells. In relation to rates of molecular evolution it is germline mutations that are of interest. Whatever conclusions can be drawn from these studies need to be extrapolated to germ cells if they are to have any evolutionary significance. Although oxidative damage to human sperm is known to occur,[18] the need for this extrapolation does introduce an additional source of uncertainty in the interpretation of results.

METABOLIC RATE AND EVOLUTIONARY RATE

The other source of evidence for an effect of metabolic rate on mutation rate is the comparison of rates of nucleotide substitution in ectotherms and endotherms. Martin et al[11] estimated rates of transversion substitutions, at 4-fold degenerate sites in the mitochondrial loci, cytochrome b and cytochrome oxidase I of 13 species of lamnoid and carcharinoid shark. The rates were obtained by estimating divergence times from the fossil record. We have discussed the problems with this approach and the way it has led to incorrect conclusions in mammals. The authors suggest that problems with the interpretation of shark fossils are not as great since fossil shark teeth are abundant and reasonably continuous. However, misinterpretation is possible even when the record is relatively complete. The transversion substitution rates among the sharks were estimated to be approximately seven times lower than those at the ND4 and ND5 genes of primates and artiodactyls. This study also involved an estimation of the rate of substitution of transitions in the cytochrome b locus between individuals in two bonnethead shark populations recently (3.5 Ma) separated by the Isthmus of Panama. The estimated rate is 2- to 3-fold lower than that estimated for primates and artiodactyls.

This estimated difference in rate between sharks and mammals is not explained by difference in generation time, since the generation time of sharks is approximately the same as that of primates and ungulates. The authors consider two other explanations: difference in DNA repair efficiency, and difference in metabolic rate. In favoring the latter they draw on the studies discussed above on oxidative damage to DNA.

Slower rates of nucleotide substitution in ectotherms have also been inferred from comparative analyses of mtDNA restriction fragments[5] (Table 12.1) and nucleotide sequences variation at the cytochrome b locus[4] of Testudine turtles. Substitution rates were estimated from divergence times derived from biogeographical as well as fossil evidence. For nonmarine species divergence times are poorly estimated, and in two cases the estimates are made by assuming that they occurred at the same time as the divergence of nonturtle species in the same geographic localities. For the Pacific and Atlantic populations of the Green turtle and the Kemp's and olive ridleys, rates are based on the time since the formation of the Panamanian Isthmus. As the authors note and later reiterate,[4] however, all the marine species sampled are highly mobile and migratory and the hypothesis of lower rates of sequence

Table 12.1. Estimated magnitudes of the "slowdown" in sequence divergence for turtle mtDNA

Comparison	Divergence Time (Ma)	Sequence Divergence	Substitution Rate Slowdown[c]
Green turtle, Atlantic - Pacific	3.0[a]	0.6	10-fold
Olive Ridley - Kemp's Ridley	3.0[a]	1.2	5-fold
Diamond-back terrapin, Atlantic - Gulf Coast	0.7[b]	0.1	14-fold
Slider, Eastern - Western	3.0[b]	0.6	10-fold
Desert Tortoise, Eastern - Western	5.5[b]	5.3	2-fold

a. calculated from geological evidence
b. calculated from comparisons with variation in mtDNA of nonturtle species
c. relative to rate observed in mammals
Reproduced with permission from JC Avise, Mol Biol Evol 1992; 9:457-473.

divergence could easily be compromised by long-range movement. Avise et al,[5] while accepting the inadequacies of their substitution rate estimates, point out that all of their independent estimates are substantially lower than mammalian estimates, suggesting that there is some real difference.

Martin and Palumbi[12] compare mitochondrial genome substitution rates among a range of endothermic and ectothermic vertebrates. In their analysis molecular distances are based on restriction fragment data, and substitution rate estimates are based on divergence times derived from fossil or biogeographical evidence. Both of these factors introduce substantial error. They derive relationships between body size (and hence metabolic rate) and substitution rate for endothermic and ectothermic vertebrate taxa (Fig. 12.2). With respect to their comparison among mammals, their conclusions depend entirely on the same kinds of assumptions about divergence times that we have discussed in previous chapters. Thus, for example, they assume a 10–12 Ma divergence of mice and rats, which we have shown is extremely improbable.

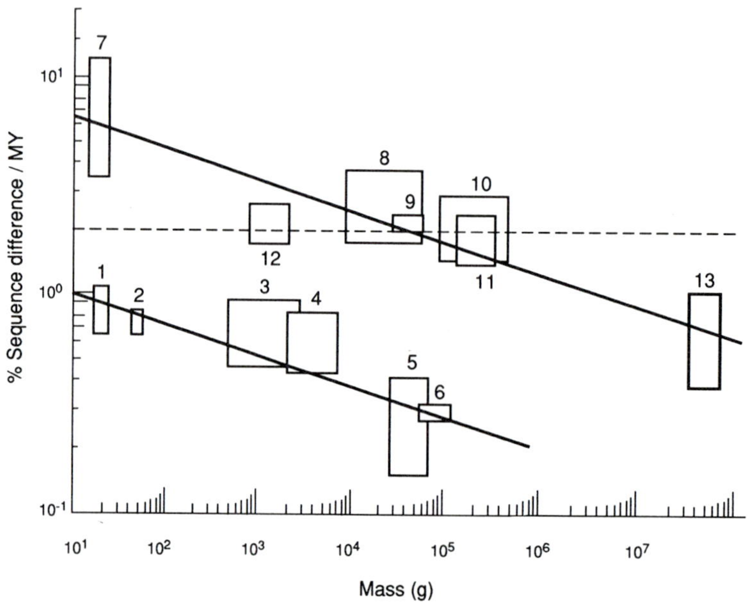

Fig. 12.2. Proposed relationship between the rate of mitochodrial DNA sequence divergence, body size and metabolic rate. The area of the boxes represents the ranges of substitution rate and body mass for a given taxon. Taxa are: 1, newts; 2, frogs; 3, tortoise; 4, salmon; 5, marine turtles; 6, sharks; 7, mice; 8, dogs; 9, primates; 10, horses; 11, bears; 12, geese; 13, whales. Modified and reproduced with permission from AP Martin and SR Palumbi, Proc Natl Acad Sci (USA) 1993; 90:4087-4091.

Adachi et al[3] compared amino acid sequences derived from the nucleotide sequences of entire mitochondrial genomes of five mammals, a bird, an amphibian and a fish. Their phylogenetic analysis indicates the following order of substitution rates: mammal > bird > amphibian > fish, there being a 6- to 10-fold difference in rate between mammals and fish. They interpret these differences as being due to variation in the degree of selective constraint on proteins between ectotherms and endotherms (also suggested previously by Thomas and Beckenbach[19]) and to variation in mutation rate. They do not attempt to distinguish among cell-generation time, metabolic rate and other possible causes for the latter.

Attempts to identify an effect of metabolic rate on mutation rate that depend on comparisons among different vertebrate classes will always be compromised by the many other factors that differ between classes. A better approach is to compare closely related ectothermic and endothermic species. In fish, endothermy has arisen more than once among the large oceanic mackerels, tunas and billfishes. Comparison of nucleotide sequences for the cytochrome b gene among these fishes,[20] including both endothermic and ectothermic species, clearly shows that there is no increased rate of substitution in the endothermic species.

SUMMARY

Evidence that metabolic rate affects mutation rate comes from studies in which rates of oxidative damage of DNA are compared among species with different metabolic rates, and from comparative analysis of nucleotide sequences in ectothermic and endothermic species. Evidence from the first source is indirect, inconclusive and far from convincing. Evidence from the second source suggests that substitution rates in the mitochondrial genome may have been greater in endothermic than in ectothermic vertebrates. For the most part this evidence depends on estimates of divergence times derived from fossil or biogeographical evidence and is, as we have argued elsewhere in this book, therefore inherently suspect. Despite this there does appear to be sufficient evidence to accept a difference in rate among vertebrate classes. However, this conclusion is based on comparisons between widely divergent taxa (e.g. mammals and turtles) which differ in many more respects than just their metabolic rates. If substitution rate differences do exist between these taxa, these may be due to other factors. In this context it should be noted that analysis thus far has focused on the mitochondrial genome. There is no evidence of rate difference in the nuclear genome, and thus the cause of rate differences may be specific to the mitochondrial genome. Vawter and Brown[21] showed that, whereas the mitochondrial genome evolves much faster than the nuclear genome in mammals, this does not appear to be the case in insects.

The one comparison of sequences among vertebrate classes in which the relative rate approach was used,[3] suggests that metabolic rate may not be the cause. It indicates a substantially higher rate of substitution in mammals than in birds, which is directly contrary to the metabolic rate hypothesis, as birds have higher body temperatures and higher metabolic rates than comparably sized mammals. It also suggests a substantial difference in substitution rate between fish and frogs, which are both ectotherms. The apparent lack of difference in substitution rate between closely related ectothermic and endothermic fish species further suggests that metabolic rate does not have any appreciable effect on mutation rate.

There does appear to be substitution rate variation in mitochondrial genes among vertebrate classes, but the degree of this variation is not clear. To the extent that estimating the degree of variation depends on fossil interpretation, it may never be clear. It is also not clear what causes the variation. There are a number of inconsistencies with a simple metabolic rate hypothesis. Certainly, within mammals the results of relative rate tests show that the metabolic rate difference that exists between rodents and primates does not affect substitution rate. We have discussed the limitations of the evidence for a relationship between metabolic rate and oxidative damage to DNA. It is not clear on current evidence that the relationship exists. If it does, it would appear, at least in mammals, that there is not a simple relationship between the degree of oxidative damage and mutation rate. Either way, there are important implications for understanding the mechanisms underlying the degenerative processes that are attributed to oxidative DNA damage.

References

1. Rand DM. Thermal habit, metabolic rate and the evolution of mitochondrial DNA. Trends Ecol Evol 1994; 9:125-131.
2. Martin AP, Naylor GJP, Palumbi SR. Rates of mitochondrial DNA evolution in sharks are slow compared with mammals. Nature 1992; 357:153-155.
3. Adachi J, Cao Y, Hasegawa M. Tempo and mode of mitochondrial DNA evolution in vertebrates at the amino acid sequence level: rapid evolution in warm-blooded vertebrates. J Mol Evol 1993; 36:270-281.
4. Bowen BW, Nelson WS, Avise JC. A molecular phylogeny for marine turtles: trait mapping, rate assessment, and conservation relevance. Proc Natl Acad Sci (USA) 1993; 90:5574-5577.
5. Avise JC, Bowen BW, Lamb T, Meylan AB, Bermingham E. Mitochondrial DNA evolution at a turtle's pace: evidence of low genetic variability and reduced microevolutionary rate in the testudines. Mol Biol Evol 1992; 9:457-473.
6. Ames BN. Endogenous DNA damage as related to cancer and aging. Mut Res 1989; 214:41-46.
7. Loeb LA. Endogenous carcinogenesis: molecular oncology into the twenty-first century—presidential address. Cancer Res 1989; 49:5489-5496.

8. Ames BN, Shigenaga MK, Hagen TM. Oxidants, antioxidants, and the degenerative diseases of aging. Proc Natl Acad Sci (USA) 1993; 90:7915-7922.
9. Larsen PL. Aging and resistance to oxidative damage in *Caenorhabditis elegans*. Proc Natl Acad Sci (USA) 1993; 90:8905-8909.
10. von Sonntag C. The chemical basis of radiation biology. London: Taylor and Francis, 1987.
11. Martin AP, Naylor GJ, Palumbi SR. Rates of mitochondrial DNA evolution in sharks are slow compared with mammals. Nature 1992; 357:153-155.
12. Martin AP, Palumbi SR. Body size, metabolic rate, generation time, and the molecular clock. Proc Natl Acad Sci (USA) 1993; 90:4087-4091.
13. Cathcart R, Schwiers E, Saul RL, Ames BN. Thymine glycol and thymidine glycol in human and rat urine: a possible assay for oxidative DNA damage. Proc Natl Acad Sci (USA) 1984; 81:5633-5637.
14. Adelman R, Saul RL, Ames BN. Oxidative damage to DNA: relation to species metabolic rate and life span. Proc Natl Acad Sci USA 1988; 85:2706-2708.
15. Shigenaga MK, Gimeno CJ, Ames BN. Urinary 8-hydroxy-2'-deoxyguanosine as a biological marker in in vivo oxidative DNA damage. Proc Natl Acad Sci (USA) 1989; 86:9697-9701.
16. Fraga CG, Shigenaga MK, Park J, Degan P, Ames BN. Oxidative damage to DNA during aging: 8-Hydroxy-2'-deoxyguanosine in rat organ DNA and urine. Proc Natl Acad Sci (USA) 1990; 87:4533-4537.
17. Park E, Shigenaga MK, Degan P, Korn TS, Kitzler JW, et al. Assay of excised oxidative DNA lesions: Isolation of 8-oxoguanine and its nucleoside derivatives from biological fluids with a monoclonal antibody column. Proc Natl Acad Sci (USA) 1992; 89:3375-3379.
18. Fraga CG, Motchnik PA, Shigenaga MK, Helbock HJ, Jacob RA, et al. Ascorbic acid protects against endogenous oxidative DNA damage in human sperm. Proc Natl Acad Sci (USA) 1991; 88:11003-11006.
19. Thomas WK, Beckenbach AT. Variation in salmonid mitochondrial DNA: evolutionary constraints and mechanisms of substitution. J Mol Evol 1989; 29:233-245.
20. Block BA, Finnerty JR, Stewart AFR, Kidd J. Evolution of endothermy in fish: mapping physiological traits on a molecular phylogeny. Science 1993; 260:210-214.
21. Vawter L, Brown WM. Nuclear and mitochondrial DNA comparisons reveal extreme rate variation in the molecular clock. Science 1986; 234:194-196.

CHAPTER 13

SUMMARY

Our aim in this book has been to present and evaluate the evidence relating to the molecular clock in mammals. Because we have focused on this single issue, there are many related topics that we have dealt with either superficially or not at all. Our concern has been with the comparison of evolutionary rates among different lineages.

In this context we have provided relevant information about the paleontology and biogeography of mammalian subclasses and placental orders, especially in primates. We have given little attention to the issue of variation in evolutionary rates among different regions or components of the genome, for which there is now extensive evidence.[1,2] We have also not dealt with the genomic factors that might be the cause of such variation. These include: codon usage bias, neighboring base composition, regional base composition and transcriptional selection.[3-6] We have also provided no detailed discussion of the numerous statistical procedures used in molecular evolutionary studies, including those for correcting for multiple substitutions in sequence data, and for estimating phylogenies.[7] Finally we have not dealt with the complex issues relating to rates of molecular evolution within species. It is not that any of these topics are unimportant or uninteresting, but this book is not intended as a comprehensive text on molecular evolution and many of the topics have been dealt with more than adequately elsewhere.

In tracing the historical development of studies on the molecular clock in mammals we have distinguished between those studies in which conclusions have depended on fossil interpretation, and those in which they have not. When these two approaches provided different results in early studies of primates this led to extensive reinterpretation of the fossil evidence. Our analysis indicates that further revision is required within primates, consistent with a much later divergence of humans from other taxa. The analysis also indicates the need for revision of the dates of separation of placental orders and of placentals from marsupials.

Currently comparative sequence analysis in mammals (or any other taxon) is limited by the availability of extensive sequence data for homologous genome regions in only a very small number of species. This should change in the near future as the ease of obtaining suitable sequence data increases and the cost of obtaining it decreases. The need to revise divergence times that we have identified is evident from analysis of only a small number of taxa. As sequence comparisons include more species it is likely that our understanding of the pattern of mammalian evolution will need ongoing revision.

In chapter 10 we obliquely raised the related issue of the role of molecular data in taxonomy (as opposed to systematics). We suggested that the very small genetic distances among African apes warranted their inclusion in the same genus (*Homo*). In doing this we assumed that molecular distances are a valid basis for taxonomic categorization. The appropriateness of this assumption depends on the purpose of the categorization. At present the decision as to whether species should be grouped into genera or families or orders etc. is arbitrary and is made on the basis of consensus, reflecting broadly held views about the relative morphological similarity of different species. Perceptions of morphological similarity are inevitably subjective to some degree, and morphological variation is not always related to genetic variation. The current approach is thus little more than a convenient way of naming and ordering species.

Perhaps the most important contribution of molecular data to evolutionary biology has been that they allow the separate investigation of evolutionary pattern and evolutionary process. Molecular data give an understanding of pattern, i.e., the order and timing of species' separations, and this provides a framework for the analysis of the processes of speciation and morphological change. Modern taxonomy aims to represent the pattern rather than the process of evolutionary change. Defining taxonomic ranks in terms of molecular distances would be an appropriate reflection of the important role of molecular data in understanding the pattern of evidence. It would enable organisms to be classified in a less arbitrary way than they presently are.

We have mentioned the experimental results that show that mutation rates are approximately the same in tissues in which cell division occurs at very different rates. This is consistent with our finding of rate uniformity among mammalian lineages with different cell generation times. The implication of this lack of difference in rate is that the mutations that are substituted in evolution do not arise as errors in DNA replication. There is some evidence that rates of substitution on Y-chromosomes are faster than rates on autosomes which are faster than rates on X-chromosomes, although the evidence is far from conclusive. This is expected if there is a cell generation time effect on substitution rate, with a greater number of cell generations occurring in males than in females. It is also expected, however, as a result of other differences between male and female germ lines. We have discussed two such differences: the differential occurrence of methylation, and the smaller effective population size of Y-chromosomes. Thus any finding of a sex difference in substitution rate is not necessarily inconsistent with the absence of a cell-generation time effect.

The uniformity of substitution rates in humans and rodents, which differ substantially in their metabolic rates, indicates that, within mammals, metabolic rate does not affect substitution rate. There is inconclusive evidence of an effect of metabolic rate on the rate of occurrence of DNA lesions, but no direct evidence of an effect on the rate of mutation. Differences in substitution rate between endothermic and exothermic species in different vertebrate classes have been interpreted as demonstrating that metabolic rate affects mutation rate. However, the species being compared are distantly related and differ from each other in a variety of ways. It seems inappropriate to single out thermal physiology as the only one among many differences that may cause a difference in substitution rate.

The molecular clock is an important concept with important implications, and it has been the subject of intense controversy over a long period of time. We contend that much of this controversy has arisen from inadequate or inappropriate interpretation of data. We hope that, in addition to raising important issues about mammalian evolution, we have made some contribution to developing a more critical investigative approach.

References

1. Wolfe KH, Sharp PM, Li W-H. Mutation rates differ among regions of the mammalian genome. Nature 1989; 337:283-285.
2. Wolfe KH, Sharp PM. Mammalian gene evolution: Nucleotide sequence divergence between mouse and rat. J Mol Evol 1993; 37:441-456.
3. Bernardi G, Olofsson B, Filipski J, Zerial M, Salinas J, et al. The mosaic genome of warm-blooded vertebrates. Science 1985; 228:953-958.
4. Ticher A, Graur D. Nucleic acid composition, codon usage, and the rate of synonymous substitution in protein-coding genes. J Mol Evol 1989; 28:286-298.
5. Yomo T, Ohno S. Concordant evolution of coding and noncoding regions of DNA made possible by the universal rule of TA/CG deficiency—TG/CT excess. Proc Natl Acad Sci (USA) 1989; 86:8452-8456.
6. Sharp PM, Stenico M, Peden JF, Lloyd AT. Codon usage: mutational bias, translational selection, or both? Biochem Soc Trans 1993; 21:835-841.
7. Nei M. Molecular evolutionary genetics. New York: Columbia University Press, 1987.

GLOSSARY

African apes: in this book, comprises gorilla, both species of chimpanzee *and* humans; recognized as close relatives since the 1860s. African apes are the sister group to the orangutan (*Pongo*).

Agnatha: the vertebrate class of "jawless" fishes; e.g. the lamprey; living agnathans are the sister group to all other vertebrates.

angiosperms: flowering plants.

apes: an informal synonym for the Hominoidea. The early crude definition was any monkey-like creature without a tail; this still persists in some common names such as "Barbary ape" and the "Celebes black ape", both of which are in fact Old World monkeys (Cercopithecoidea).

archosaurs: dinosaurs and their relatives. Two groups survive: crocodilians and birds.

artiodactyl: informal name for members of the Artiodactyla, the "even-toed" ungulates.

autosome: any chromosome that is not a sex chromosome; one of the set of homologous chromosome pairs that is common to both sexes.

Beringia: the land bridge presently submerged beneath the Bering Straits, between Alaska and Siberia.

bottleneck: a short-term reduction in the size of a population, during which allele frequencies may change more rapidly than usual due to genetic drift.

bovid: informal name for members of the Bovidae, a diverse artiodactyl family of browsers and grazers; cattle, bison, sheep, goats, antelope and gazelles.

carnivore: informal name for members of the Carnivora; specialized medium-to-large meat-eating placentals The order is divided into the Feliformia (cats, hyaenas etc.) and the Caniformia (dogs and bears and their relatives).

catarrhine: informal name for members of the Catarrhini, the Old World monkeys and apes (including humans).

caviomorph: informal name for members of the Caviomorpha, a rodent group peculiar to the Americas, mainly South America; named after *Cavia*, the guinea pig.

Cenozoic Era: the "age of mammals", from 65 Ma ago to the present; divided into the Tertiary (till 1.64 Ma ago) and the Quaternary (1.64 Ma to present) sub-eras; divided into 7 epochs and into the Paleogene and Neogene periods.

cercopithecoid: informal name for members of the Cercopithecoidea, the Old World monkeys.

Cetacea: whales and dolphins, probably related to artiodactyls.

chimpanzee: comprises at least two species, *Pan troglodytes*, the common chimpanzee, and *Pan paniscus*, the bonobo or "pigmy" chimpanzee (which is not much smaller). *Pan* is the living sister group to *Homo*.

Chiroptera: almost a thousand living species of bats, commonly divided into megabats (Megachiroptera) and microbats (Microchiroptera).

clade: a "branch" of a phylogenetic tree; a taxon or other group that is monophyletic. See cladistics.

cladistics: a method of systematics and taxonomy based on the work of Willi Hennig, a German taxonomist, who realized that synapomorphies were required to diagnose monophyletic groups (clades); groups diagnosed by symplesiomorphies were paraphyletic.

Cretaceous Period: from 145.6 to 65 Ma ago.

Cricetidae: a diverse family of rodents comprising voles, lemmings, gerbils and hamsters; thought to be the paraphyletic sister group to murids.

dentary-squamosal articulation: type of jaw joint that is peculiar to mammals. The squamosal is a skull bone; a dentary is a lower jaw bone that bears teeth. In mammals a mandible comprises two dentaries only, whereas in other jawed vertebrates lower jaws include other bones.

dermopteran: informal name for members of the Dermoptera, the colugos, or "flying lemurs" from Southeast Asia.

diploid: describes a cell with one set of pairs of homologous chromosomes; in mammals most cells except gametes are diploid. Contrast with haploid.

edentate: informal term for Xenarthra and Pholidota. The formal "Edentata" has been abandoned as edentates are collectively paraphyletic or polyphyletic.

effective population size: a statistically ideal population in which all individuals of either sex have an equal chance of being a parent and hence of passing on genes to the next generation. The effective population size may be much smaller than the real population size.

Eocene Epoch: 56.5 Ma to 35.4 Ma ago.

generation time: the mean time from conception to breeding.

genetic drift: changes in gene frequency that occur due to random sampling of the alleles that are passed on from one generation to the next.

genetic marker: a polymorphic gene or other DNA sequence used to study individual variation.

gibbons: informal name for Hylobatidae, the lesser apes of Southeast Asia, the sister group to the great apes (including humans).

Gondwana: the ancient southern supercontinent that at various times included Africa, South America, Antarctica, Australia, New Zealand, Madagascar and India.

gorilla: three subspecies of *Gorilla gorilla*, the endangered sister group to the chimpanzee-human clade.

haploid: haploid cells contain a set of only one of each pair of homologous chromosomes; gametes are normally haploid. Contrast with diploid.

Haplorhini: comprise *Tarsius* and its sister group, the simians (Simiiformes).

heterozygote: individual or cell with different alleles at the same locus on homologous chromosomes. Contrast with homozygote.

Holarctic region: large northern biogeographic region divided into the Nearctic Region (North America), and the Palearctic Region (Europe, Asia north and west of the Himalayas, Africa north of the Sahara, and the Arabian Peninsula except for its southeast coastline).

Holocene Epoch: the last 10,000 years of geological time.

hominoid: informal name for members of the Hominoidea (the apes) comprising the gibbons (Hylobatidae) and the great apes; hominoids are the sister group of the Cercopithecoidea.

homologous: means "corresponding", "having the same relation". See homology. When referring to chromosomes: a pair of chromosomes that generally have the same loci in the same positions.

homology: similarity due to putative inheritance from a shared common ancestor. Contrast with homoplasy. See orthologous and paralogous.

homoplasy: an apparently shared character that is actually due to parallel or convergent evolution. For example, wings have been evolved at least three times among tetrapods (archosaurs, birds, bats); though the forelimb bones that support the wings have homologies in all three groups, the wings themselves are homoplasies.

homozygote: an individual with the same allele at the same locus on homologous chromosomes. Contrast with heterozygote.

Hyracoidea: hyraxes or conies, from Africa and Western Asia.

hystricognath: informal name for members of the Hystricognatha, a suborder of Rodentia; hysticognaths comprise the Caviomorpha, the Old World porcupine *Hystrix* and its relatives (Hystricidae) and some unusual African families of rodents. The non-caviomorph hystricognaths are sometimes called the Phiomorpha, which may be paraphyletic.

insectivore: informal name for Lipotyphla and Macroscelidea; the formal name "Insectivores" has been abandoned as such a group is paraphyleytic or polyphyletic.

Jurassic Period: from 208 to 145.6 Ma ago.

lagomorph: informal name for members of the Lagomorpha: rabbits, hares and pikas.

Laurasia: the ancient northern supercontinent, comprising Europe, North America and Greenland, and most of Asia.

Lipotyphla: insectivores such as moles and porcupines.

locus (plural: loci): the particular site on a chromosome at which a gene or genetic marker is located, hence refers to a gene with all its variant alleles.

Macroscelidea: the elephant shrews of Africa.

Marsupialia: formal name for the clade of pouched mammals such as kangaroos and koalas. This term has clear priority over its junior synonym, Metatheria. Generally found in Australia, New Guinea and South America, a well-known exception is the American opossum *Didelphis virginianis*. Marsupials are so named because they possess pouches; however this does not apply to males nor even the females of all species.

Mesozoic Era: from about 245 Ma ago to 65 Ma ago; comprises the Triassic, Jurassic and Cretaceous periods.

Miocene Epoch: 23.3 Ma to 5.2 Ma ago.

molecular clock: the subject of this book. The theory that substitutions in proteins and DNA occur at a stochastically uniform rate over time, hence the genetic distance between taxa is generally proportional to the time elapsed since their last common ancestor.

monophyletic: among living taxa, a monophyletic group (or clade) includes all the living descendants of a presumed common ancestor (e.g. to be monophyletic, great apes must include humans). If paleospecies are included, then a monophyletic group includes the presumed common ancestor and all of its the descendants. See cladistics.

monotreme: informal name for members of the Monotremata, the clade of egg-laying mammals, comprising the platypus and two species of echidna. Monotremes are sometimes placed within Prototheria.

murids: informal name for members of the Muridae, the family of rodents that includes mice and rats (Murinae). Living murids comprise over 100 genera and many hundreds of species. There are three related living families: the Spalacidae (Old World "vole rats"), the Rhizomyidae ("bamboo rats"), and the Cricetidae (hamsters etc.); the latter are thought to be the sister group of the Muridae.

Murinae: the subfamily of murids that includes mice and rats.

Neogene Period: from 23.3 to 1.64 Ma ago; comprises the Miocene and Pliocene Epochs.

New World monkeys: informal name for the Platyrrhini.

Old World monkeys: informal name for the Cercopithecoidea.

Oligocene Epoch: 35.4 Ma to 23.3 Ma ago.

orangutan: two subspecies of *Pongo pygmaeus*, the great ape of Southeast Asia, confined respectively to Borneo and to Sumatra.

orthology: homology arising by speciation. Contrast with paralogy.

paleospecies: an extinct species known only from fossils.

Paleocene Epoch: 65 Ma to 56.5 Ma ago.

Paleogene: from 65 to 23.3 Ma ago; comprises Paleocene, Eocene and Oligocene Epochs.

Pangea: the ancient supercontinents that have included all major land masses. Pangea repeatedly forms and breaks up over a 400-500 Ma cycle.

paralogy: homology arising by gene duplication. Contrast with orthology.

paraphyletic: a group of common descent that does not include a group of the known descendants. For example, "dinosaurs" as commonly recognized are paraphyletic, as that name does not usually include birds, which are descended from them. Contrast with monophyletic. See cladistic.

pecoran: informal name for members of the Pecora, generally horned ruminants, the most diverse artiodactyl clade; pecorans comprise the Bovidae, Giraffidae (giraffes and the okapi), Cervidae (deer) and Antilocapridae (North American prong-buck).

Perissodactyla: "odd-toed" ungulates such as horses.

pholidote: informal name for members of the Pholidota, pangolins of Africa and South Asia.

phylogeny: the evolutionary history of a taxon, often shown diagramatically as a phylogenetic tree.

placental or placental mammal: the Placentalia. Some prefer to describe these mammals as Eutheria or eutherians, because marsupials also possess placentas. However, Placentalia has formal taxonomic priority.

platyrrhine: informal name for members of the Platyrrhini, all primates in South and Central America, placed in about 16 genera.

Pleistocene Epoch: from 1.64 Ma to 0.01 Ma ago; includes the recent "ice ages".

Pliocene Epoch: from 5.2 Ma to 1.64 Ma ago.

polyphyletic: an unnatural grouping of organisms that excludes more than one group of descendants from various common ancestors. For example, grouping hyaenas with dogs would be polyphyletic, because hyaenas share a common ancestor with cats while dogs share a common ancestor with bears. See cladistics.

Primates: about 180 species of monkeys and their relatives, ranging from humans to bush babies. Primates are divided into two sub-orders, the Strepsirhini and the Haplorhini.

Proboscidea: two species of elephants.

Prototheria: a paraphyletic or polyphyletic taxon that includes those Mesozoic mammals lacking the characteristics of therians.

Quaternary Sub-era: from 1.64 Ma ago till the present.

Rodentia: rodents, comprising rats and mice (Murinae) and all their many relatives such as squirrels and beavers. Rodents have as many species as all other placental orders combined.

scandentian: informal name for members of the Scandentia: tree shrews.

sex chromosomes: a special pair of chromosomes that are not fully homologous and that are involved in determining sex (but not in all animals). Where present, sex chromosomes do not occur in equal number in the cells of males and females. In mammals, females' cells have two X chromosomes and males' cells have an X and a Y chromosome.

simians: informal name for the Simiiformes, the monkeys and apes, divided into Platyrrhini (New World Monkeys) and the Catarrhini (Old World monkeys and apes); living simians are the sister group of *Tarsius*.

Sirenia: "sea cows" or dugongs, and manatees.

sister group: if two taxa are sister groups, then they are each other's closest relatives and together form a monophyletic group or clade.

Strepsirhini: comprise two biogeographic groups, the lemurs of Madagascar, and the less diverse but more numerous lorises and galagos (bush babies) of Africa and Asia. Strepsirhines are the sister group of the Haplorhini.

symplesiomorphy: a shared ancestral character. For example, feathers are a symplesiomorphy among birds: all birds have feathers so this is no use for systematics or taxonomy within the birds. See cladistics.

synapomorphy: a shared derived character that is possessed by only some of the taxa being compared. The implicit hypothesis is that of presumed descent from a common ancestor. For example, among the tetrapods, only birds have feathers, by reason of their presumed common descent from a unique shared common ancestor that also had feathers. See cladistics.

Synapsida: an ancient group of "reptiles" that include the ancestors of modern mammals. In traditional paleontolgical use, the taxon is paraphyletic.

systematics: the study of the evolutionary relationship of taxa, both living and extinct.

Tarsius: four species of small active nocturnal and carnivorous haplorhines from Southeast Asia; the sister group of simians.

taxon (plural taxa): any group of organisms that has a formal Linnean name. In zoology, the lowest taxonomic category is the subspecies.

taxonomy: in the recent narrow sense, the formal rules for classifying taxa; in the older wider sense, included systematics as well.

Tertiary Sub-era: from 65 to 1.64 Ma ago.

Tethys Sea: the ancient sea that divided Laurasia from Gondwana; remnants remain such as the Mediterranean and the Black Sea.

tetrapods: four-legged vertebrates.

Theria: marsupials and placentals, formally a subclass of Mammalia. Therians are widely held to be monophyletic, and include the living tribosphenids.

transgenic: an individual that expresses a gene that has been artificially introduced from another species.

Triassic Period: from 245 to 208 Ma ago.

tribosphenids: mammals with "tribosphenic" molars. Tribosphenic molars have a mortar-and-pestle action; the upper molars have an elaboration of the protocone which matches a talonid basin in lower molars.

Tubulidentata: the peculiar aardvark of Africa has its own order.

Turgai Sea: an ancient epicontinental (over continental crust) sea that lay to the east of the present-day Urals; a northerly branch of the Tethys. The Turgai divided Asia from Europe for much of the Paleogene.

ungulates: hoofed placental mammals and their relatives. Of living placentals, these obviously include the Artiodactyla and Perissodactyla. However, the Cetacea, Hyracoidea, Proboscidea and Sirenia may also be related; if so, hooves have been lost secondarily in these latter orders.

vertebrate: informal name for members of the Vertebrata, animals with backbones; sub-phylum of the chordates (Chordata).

vicariance: in biogeography, the opposite of dispersion. A vicariant distribution arises when related species are passively separated due to geographic processes, such as changes in sea level or continental drift. For example, the occurrence of marsupials in both Australia and South America is vicariant, reflecting the ancient breakup of Gondwana, as opposed to a later dispersion across the Pacific Ocean.

xenarthran: informal name for members of the Xenarthra, the armadillos, sloths and anteaters of the Americas. The name refers to a peculiarity of the joints between their vertebrae.

INDEX

Page numbers in italics denote figures (f) and tables (t).

N

O

P

R

S

T

V

W

X

Y

Z

Molecular Biology Intelligence Unit

Available and Upcoming Titles

- ❐ Organellar Proton-ATPases
  Nathan Nelson, Roche Institute of Molecular Biology
- ❐ Interleukin-10
  Jan DeVries and René de Waal Malefyt, DNAX
- ❐ Collagen Gene Regulation in the Myocardium
  M. Eghbali-Webb, Yale University
- ❐ DNA and Nucleoprotein Structure In Vivo
  Hanspeter Saluz and Karin Wiebauer, HK Institut-Jena and GenZentrum-Martinsried/Munich
- ❐ G Protein-Coupled Receptors
  Tiina Iismaa, Trevor Biden, John Shine, Garvan Institute-Sydney
- ❐ Viroceptors, Virokines and Related Immune Modulators Encoded by DNA Viruses
  Grant McFadden, University of Alberta
- ❐ Bispecific Antibodies
  Michael W. Fanger, Dartmouth Medical School
- ❐ Drosophila Retrotransposons
  Irina Arkhipova, Harvard University and Nataliya V. Lyubomirskaya, Engelhardt Institute of Molecular Biology-Moscow
- ❐ The Molecular Clock in Mammals
  Simon Easteal, Chris Collet, David Betty, Australian National University and CSIRO Division of Wildlife and Ecology
- ❐ Wound Repair, Regeneration and Artificial Tissues
  David L. Stocum, Indiana University-Purdue University
- ❐ Pre-mRNA Processing
  Angus I. Lamond, European Molecular Biology Laboratory
- ❐ Intermediate Filament Structure
  David A.D. Parry and Peter M. Steinert, Massey University-New Zealand and National Institutes of Health
- ❐ Fetuin
  K.M. Dziegielewska and W.M. Brown, University of Tasmania
- ❐ Drosophila Genome Map: A Practical Guide
  Daniel Hartl and Elena R. Lozovskaya, Harvard University
- ❐ Mammalian Sex Chromosomes and Sex-Determining Genes
  Jennifer A. Marshall-Graves and Andrew Sinclair, La Trobe University-Melbourne and Royal Children's Hospital-Melbourne
- ❐ Regulation of Gene Expression in *E. coli*
  E.C.C. Lin, Harvard University
- ❐ Muscarinic Acetylcholine Receptors
  Jürgen Wess, National Institutes of Health
- ❐ Regulation of Glucokinase in Liver Metabolism
  Maria Luz Cardenas, CNRS-Laboratoire de Chimie Bactérienne-Marseille
- ❐ Transcriptional Regulation of Interferon-γ
  Ganes C. Sen and Richard Ransohoff, Cleveland Clinic
- ❐ Fourier Transform Infrared Spectroscopy and Protein Structure
  P.I. Haris and D. Chapman, Royal Free Hospital-London
- ❐ Bone Formation and Repair: Cellular and Molecular Basis
  Vicki Rosen and R. Scott Thies, Genetics Institute, Inc.-Cambridge
- ❐ Mechanisms of DNA Repair
  Jean-Michel Vos, University of North Carolina
- ❐ Short Interspersed Elements: Complex Potential and Impact on the Host Genome
  Richard J. Maraia, National Institutes of Health
- ❐ Artificial Intelligence for Predicting Secondary Structure of Proteins
  Xiru Zhang, Thinking Machines Corp-Cambridge
- ❐ Growth Hormone, Prolactin and IGF-I as Lymphohemopoietic Cytokines
  Elisabeth Hooghe-Peters and Robert Hooghe, Free University-Brussels
- ❐ Human Hematopoiesis in SCID Mice
  Maria-Grazia Roncarolo, Reiko Namikawa and Bruno Péault DNA Research Institute
- ❐ Membrane Proteases in Tissue Remodeling
  Wen-Tien Chen, Georgetown University
- ❐ Annexins
  Barbara Seaton, Boston University
- ❐ Retrotransposon Gene Therapy
  Clague P. Hodgson, Creighton University
- ❐ Polyamine Metabolism
  Robert Casero Jr, Johns Hopkins University
- ❐ Phosphatases in Cell Metabolism and Signal Transduction
  Michael W. Crowder and John Vincent, Pennsylvania State University
- ❐ Antifreeze Proteins: Properties and Functions
  Boris Rubinsky, University of California-Berkeley
- ❐ Intramolecular Chaperones and Protein Folding
  Ujwal Shinde, UMDNJ
- ❐ Thrombospondin
  Jack Lawler and Jo Adams, Harvard University
- ❐ Structure of Actin and Actin-Binding Proteins
  Andreas Bremer, Duke University
- ❐ Glucocorticoid Receptors in Leukemia Cells
  Bahiru Gametchu, Medical College of Wisconsin
- ❐ Signal Transduction Mechanisms in Cancer
  Hans Grunicke, University of Innsbruck
- ❐ Intracellular Protein Trafficking Defects in Human Disease
  Nelson Yew, Genzyme Corporation
- ❐ apoJ/Clusterin
  Judith A.K. Harmony, University of Cincinnati
- ❐ Phospholipid Transfer Proteins
  Vytas Bankaitis, University of Alabama
- ❐ Localized RNAs
  Howard Lipschitz, California Institute of Technology
- ❐ Modular Exchange Principles in Proteins
  Laszlo Patthy, Institute of Enzymology-Budapest
- ❐ Molecular Biology of Cardiac Development
  Paul Barton, National Heart and Lung Institute-London
- ❐ RANTES, *Alan M. Krensky, Stanford University*
- ❐ New Aspects of V(D)J Recombination
  Stacy Ferguson and Craig Thompson, University of Chicago

Neuroscience Intelligence Unit

Available and Upcoming Titles

- ❒ Neurodegenerative Diseases and Mitochondrial Metabolism
  M. Flint Beal, Harvard University
- ❒ Molecular and Cellular Mechanisms of Neostriatum
  Marjorie A. Ariano and D. James Surmeier, Chicago Medical School
- ❒ Ca^{2+} Regulation in Neurodegenerative Disorders
  Claus W. Heizmann and Katharin Braun, Kinderspital-Zürich
- ❒ Measuring Movement and Locomotion: From Invertebrates to Humans
  Klaus-Peter Ossenkopp, Martin Kavaliers and Paul Sanberg, University of Western Ontario and University of South Florida
- ❒ Triple Repeats in Inherited Neurologic Disease
  Henry Epstein, University of Texas-Houston
- ❒ Cholecystokinin and Anxiety
  Jacques Bradwejn, McGill University
- ❒ Neurofilament Structure and Function
  Gerry Shaw, University of Florida
- ❒ Molecular and Functional Biology of Neurotropic Factors
  Karoly Nikolics, Genentech
- ❒ Prion-related Encephalopathies: Molecular Mechanisms
  Gianluigi Forloni, Istituto di Ricerche Farmacologiche "Mario Negri"-Milan
- ❒ Neurotoxins and Ion Channels
  Alan Harvey, A.J. Anderson and E.G. Rowan, University of Strathclyde
- ❒ Analysis and Modeling of the Mammalian Cortex
  Malcolm P. Young, University of Oxford
- ❒ Free Radical Metabolism and Brain Dysfunction
  Irène Ceballos-Picot, Hôpital Necker-Paris
- ❒ Molecular Mechanisms of the Action of Benzodiazepines
  Adam Doble and Ian L. Martin, Rhône-Poulenc Rorer and University of Alberta
- ❒ Neurodevelopmental Hypothesis of Schizophrenia
  John L. Waddington and Peter Buckley, Royal College of Surgeons-Ireland
- ❒ Synaptic Plasticity in the Retina
  H.J. Wagner, Mustafa Djamgoz and Reto Weiler, University of Tübingen
- ❒ Non-classical Properties of Acetylcholine
  Margaret Appleyard, Royal Free Hospital-London
- ❒ Molecular Mechanisms of Segmental Patterning in the Vertebrate Nervous System
  David G. Wilkinson, National Institute of Medical Research, United Kingdom
- ❒ Molecular Character of Memory in the Prefrontal Cortex
  Fraser Wilson, Yale University

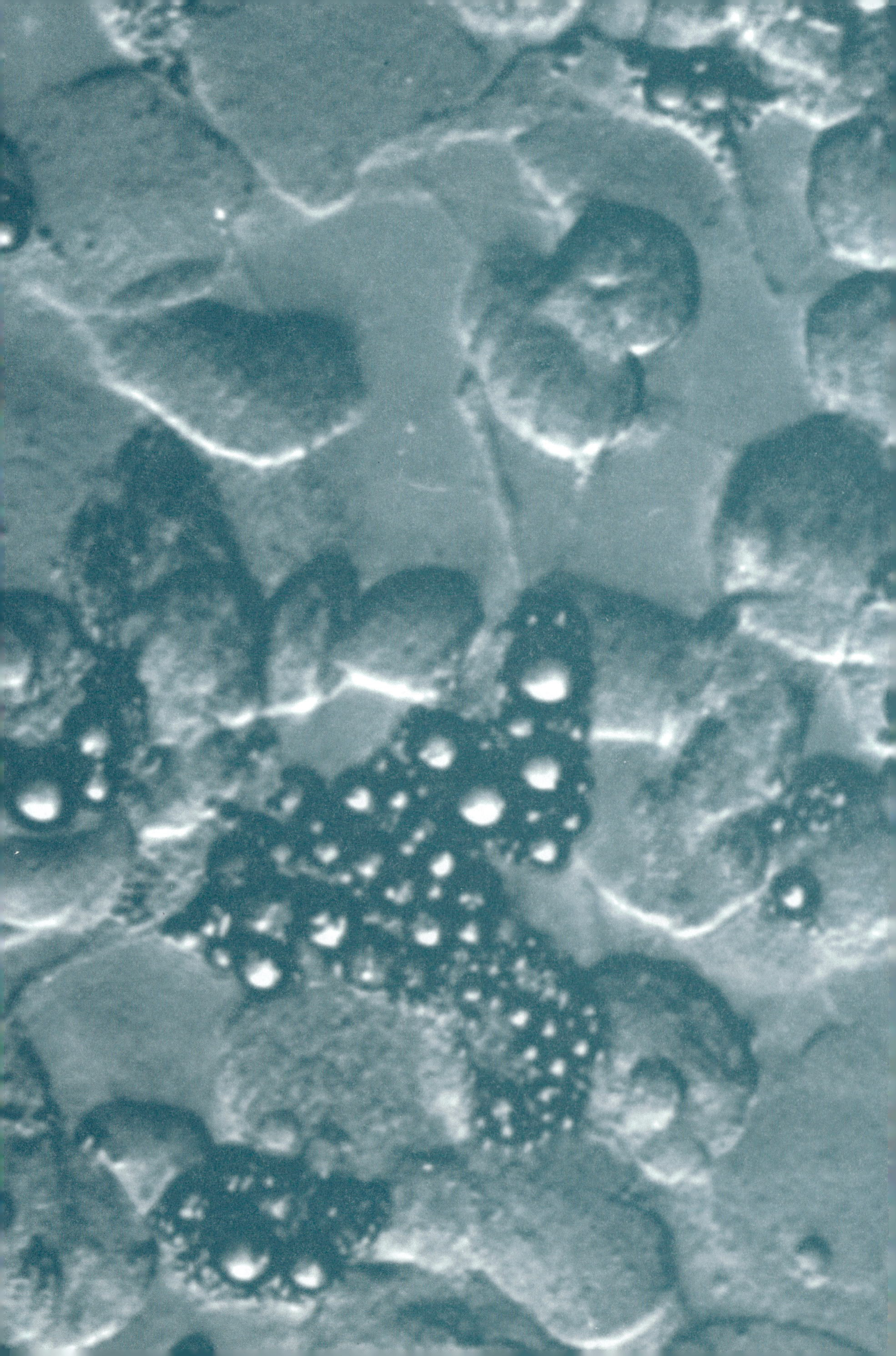